How to De-Code Equine Body Language

The Science of Horses

ISBN 978-0-9959979-2-9

The Science of Horses

Dedication

In honour and acknowledgement of all the wonderful horsemen and women who gave me access to their herds of horses and those patient and understanding owners who allowed me to sit in their pastures as I conducted my field research, studied, photographed, documented and at times conducted "experiments" to prove my equine body language theories -

this book is dedicated to you!

Table of Contents

Although the chapters follow a sequence of events, there really is no start, and certainly with a subject this complex, no end. So just dive in wherever you feel most drawn.

If you are one of those people who think

“It’s just a horse”,

this book is not for you.

Introduction

Horses have been part of man's heritage since the first guy fashioned a rope and caught himself a wild horse with the intention of putting it to work rather than eating it.

The desire to understand and communicate with horses started thousands of years ago and is just as strong today. Historically man wanted to tame and control the horse to do his will and the relationship was one of master and obedient slave.

Those days are long past and horse lovers everywhere crave a more enlightened understanding of horses. Sure, they still want the horses to be obedient and comply to their wishes, but thankfully the training equipment and methods have evolved into a more compassionate process with more emphasis on how the horse feels and responds to training and handling. This results in a more harmonious experience for both man and beast.

Warning

This book is **NOT** about horse training or natural horsemanship.

However, expanding your understanding of the horse's natural behavior and how he communicates both to other horses and humans will make all your dealings with horses much easier.

Equine Body Language

My obsession with watching horses and the desire to understand why they do the things they do started very early in life.

My first equine experiences began at age 4 when my parents enrolled me in a "learn to ride program" at the local riding school. The lessons went well and the interest in horses blossomed so they bought me a pony for my seventh birthday. It was a feisty Welsh Cob called Peggy. Despite her disagreeable personality I loved this pony with all my young heart. Peggy taught me some very important life lessons. Humility and perseverance seemed to be top of her list of things her small human should understand. Our relationship was bittersweet. As a riding pony she was frustrating and many of our rides would end with me sitting in the dirt in tears after she decided to ditch me and head back to the barn. However as a friend and confidant you could not wish for a better companion than Peggy and I spent many hours sitting in her paddock reading to her and sharing all my childhood secrets. I loved to watch her as she moved about her small field just being a pony. Peggy also seemed to enjoy this side of our relationship much better than the riding element.

Forgive me for this sappy walk down memory lane, but it was purely intended to establish a time-line for the start of this journey.

Here I am, a lifetime later, still watching horses and even more determined to understand why they do what they do.

The contents of this book will attempt to simplify a very complex subject by drawing human comparisons and using photographs to "Show and Tell" what is really going on.

Before we dive into this fascinating subject let's talk a little bit about what the term "Body Language" means, and look at some of the common myths and misconceptions.

Body language exists in both animals and humans as a means of nonverbal communication, where thoughts, intentions, or feelings are expressed by physical behaviors and actions. These can be facial expressions, body movements, gestures, body posture, eye movement, touch and the use of space.

Communicating this way is as natural as breathing and just like breathing we all communicate in a nonverbal way without even thinking about. It's as though this ability is hard wired into our DNA . With the explosion of alternative ways to communicate such as phones / text/ letters / fax/ memo /e-mail / private messaging etc. All of which do not require you to be face-to-face with the person to whom you are communicating, less and less importance has been placed on body language as a means to judge the true intent and feeling behind the words.

As far as I know horses have not mastered any of that technology which means you have to be face-to-face or at least in close proximity to be able to read their body language and likewise for them to read yours.

According to experts body language should not be confused with sign language, as sign languages are full languages with their own complex grammar systems.

Our goal is to understand how humans communicate with horses using body language.

It would be wrong not to acknowledge the extensive range of aids and cues trainers and riders have perfected over the centuries that very clearly communicate their desires to the horse. The concept of using nonverbal communication is not something new to horsemen and women. Riders use their heels and hands to control and direct the horse. They use their seat and weight to communicate speed and direction to the horse while riding. With the growing interest in Natural Horsemanship owners and trainers have taken this concept a huge step further by connecting and communicating with their horse on a much deeper level.

The focus of this book is to discover how the horse communicates, both to us and other horses using his body language and nonverbal communication. In an attempt to make this complicated subject simple and easy to understand I will set the scene so you have a better understanding of the circumstances and also draw comparisons to human behavior and how we express a similar emotion or feeling in our nonverbal communication.

As I researched into human body language I found many of the "experts" were very concerned about how to tell if someone is being truthful or lying, whether it was in a business deal or a romantic relationship. There was a lot of conflicting ideas on this topic so I came to the conclusion that there is no foolproof way to tell if someone is lying.

The good news is, when dealing with horses and reading their body language we don't need to worry about whether they are lying. Horses are always authentic. They don't pretend to be calm and relaxed when they are really stressed and anxious. They don't pretend to be lame and fake a limp to get out of work. They don't pretend to like another horse if they really don't.

But before we go any further I must mention authentic does not mean the same as honest. Confusing I know, but anyone who has ever owned a pony will attest to how sneaky and devious they can be. Ponies have an ability to scheme and plot to get what they want, whether it's a means to escape from the field and go on a little adventure or dump their rider and head off home without them.

Getting Started

Anyone can be a horse whisperer and learn to communicate with horses no matter your age or physical abilities.

The trick to learning how to read equine body language is to be still and quiet - this is not something that comes easily to humans. We like to be doing and exerting our thoughts and influence on those around us. Human beings have a difficult time just "being".
However when you start to study equine body language it is very important that you spend time as a silent observer and resist jumping to any swift judgements or conclusions.

If you have the opportunity, take a lawn chair out to the pasture where horses are living and quietly hang out for a while. On first glance it doesn't look like much is going on. The horses are either eating or snoozing but the longer you sit there the more you will become aware of their nonverbal conversations. You will be amazed at how often they touch, interact and communicate with each other. This is not an exercise you can rush so allow yourself lots of time to relax, feel at one with nature and just be.

"Horse watching" does not require any special equipment / tack or gear. All that is required is a desire to connect and an ability to quiet your spirit and really see what is going on.
You may want to take a camera and a note pad.

Linda Finstad
Fine Art Photography

Debunking a few Myths to Reveal the Truth

I am the first to admit that there is no definitive black and white answers when it comes to deciphering body language. There are many gray areas and much is subject to interpretation. However there are some very confusing myths that we need to address before we proceed.

Myths

A single action has a single specific meaning.

Understanding equine body language is a natural ability only "horse whisperers" have.

Shifty eyes are a sign of deception.

First impressions give an accurate reading of a person / horse and their personality.

You must make and keep eye contact to communicate effectively.

Body language is an unconscious act and just happens instinctively.

Nonverbal communication accounts for 93% of communication. While this number may be true for horses, humans rely heavily on the spoken word to convey information.

Fidgeting is a sign that you're nervous.

The eyes are the mirror into your soul.

Facial expressions give away your true feelings.

It takes endless studying and years of practice to be able to read body language.

It is easier to read the body language of a stranger than someone you know.

Truths

There is no one body gesture that has a single definitive meaning. You have to look at the bigger picture and the surrounding circumstances.

Reading body language is a learned skill and like most skills the more you do it the better and more accurate you become.

Shifty eyes may mean a lack of confidence and in certain cultures it is considered bad manners to make eye contact. With animals making and holding eye contact can be a confrontational gesture especially if you are staring down the alpha male.

Have you ever misjudged someone on that first meeting - only to change your opinion of them as you get to know them better? Don't feel bad it happens to us all; we are quick to judge but not always accurate in our assumptions.

Locking eyes can be taken as a sign of aggression and actively convincing rather than conveying.

As you progress through the book you will see many examples of intentional body language. However some actions are instinctive, for instance when you get a whiff of something that smells really bad, you wrinkle your nose. Likewise a stallion getting a whiff of a mare in season would instinctively curl his upper lip (flemen) to process the pheromones in the air.

Fidgeting can mean many things – or nothing at all – depending on the situation.
The most common reason for both horses and humans to fidget is because the movement takes the place of an emotion that they're unable to express, such as boredom, excitement, or frustration. Or it can simply be the by-product of excess energy from sitting or standing still for too long.

The face sometimes gives away our strongest feelings. Not true! Most adults have learned to mask their true emotions and present the face that will be most socially accepted.

It's easier to read people we are close to. You know instinctively when your spouse is ticked off, or your child is bored, or your boss wants something done, now! With people we know, we've already amassed many hours of study, and we recognize their nonverbal cues.

Old Wives' Tales

Similar to Myths - but these have been bandied about for so long they are almost considered facts. One of my personal favorites is that horses can smell fear.

Horses do have an incredible sense of smell and if in fact "fear" had a recognizable smell I am sure they would be able to detect it. However the scientists have proved sweat just smells like sweat no matter what was the cause.

Horses can tell if you are scared. True, but that has more to do with their ability to read your body language than sniff your armpits. When you are scared your breathing is shallow, and your movements are uncertain and possibly jerky. The way you walk and approach the horse will be with caution, not with calm and confidence. The horse easily reads these signs of fear and apprehension.

The best thing to do with a maiden mare who is difficult/bitchy is to breed her as it will calm her down. This may carry a grain of truth, but consider the bigger picture. The bitchy mare is now treated like a brood mare and will be allowed to simply graze in the pasture all day, which will certainly improve her disposition.

A horse has to roll COMPLETELY over to be worth a darn; the ones that roll on one side, get up and roll on other are lazy and will never amount to anything.
I feel it would be wrong to writet-off a fabulous horse just because he has a large wither and rolling completely over causes him discomfort. What do you think?

You can "teach" a horse to bite by feeding him treats.
All horses know how to bite. We don't need to teach them how to do that. What we should teach them is personal boundaries and respect. Allowing a horse to nuzzle your pockets looking for treats and then rewarded him for invading your personal space will certainly lead to someone getting a nasty bite.

Changing a horse's name is bad luck!
For whom I wonder, the owner or the horse?

Little Poems

Tradition, they say, can teach us a lot,
So here is what horsemen, on color, have thought.
A bay is hardy, a chestnut is fast
And you can't kill a buckskin: he'll last and last
A gray is gentle, a sorrel is hot.

A dun is a horse you'll be happy you bought.
White eyes are flighty, white feet may crack
While some won't rely on the feet of a black.
Some pintos are lucky, like the medicine hat,
But all horsemen agree the best color is fat.

One white foot try him.
Two white feet buy him.
Three white feet be on the sly.
Four white feet pass him by.

I am sure you have heard these pearls of wisdom and little ditties and maybe you have even found yourself repeating them without really challenging the truth behind them. I hope that after reading this book you will challenge everything, even the things you thought you knew.

The Million Dollar Question

Why do horses choose to have a relationship with us?

This question caused me the most sleepless nights.
The goal is to understand horses, not put human thoughts and feelings into their actions.
How would it be possible to unravel the mysteries of equine behavior and body language if the answer to this simple question remained unanswered?
For the longest time my conclusion was "Horses are a gift from God", which I do believe them to be. But there had to be a more scientific reason that horses would willingly befriend and serve humans in whatever way they saw fit, whether that was in the farmer's field or the battlefield. The key was to discover what could possibly motivate horses to willingly enter into such a one-sided partnership.

What Motivates a Horse?

In the quest to understand equine body language, we must first try to think the same way a horse would think. We want to discover what is important to him and learn what motivates him to do the things he does.

Although some owners refer to their horses as "fur babies" and treat them like children humans and horses are totally different species with no physiological similarities. So how can humans possibly "think like a horse"?

It turns out, the answer is not only simple but also science-based. It was there all along, in fact many of us learnt the secret to this mystery in school.
Be that as it may, for decades equine experts have overlooked the obvious connection, in search of a more complicated, or complex solution.

The answer lies in:

Maslow's Hierarchy of Needs

Abraham H. Maslow presented his findings in a scientific paper in 1943 and his theory is often represented as a pyramid with five levels of needs. This model in motivational theory, argues that while people aim to meet basic needs, they seek to meet successively higher needs to feel happy and fulfilled, once the lower needs are met. It is very easy to draw a comparison between the basic needs of humans and the basic needs of horses.

Human needs

The needs of domestic horses that are outlined on the lower three rungs of the pyramid are met by humans. This allows them to move up to the next step on the pyramid which is "seeking esteem" or love and affection from others. In the wild this need would be met from other herd members. Fortunately for man kind they turn to us to fulfill this need.

Equine needs

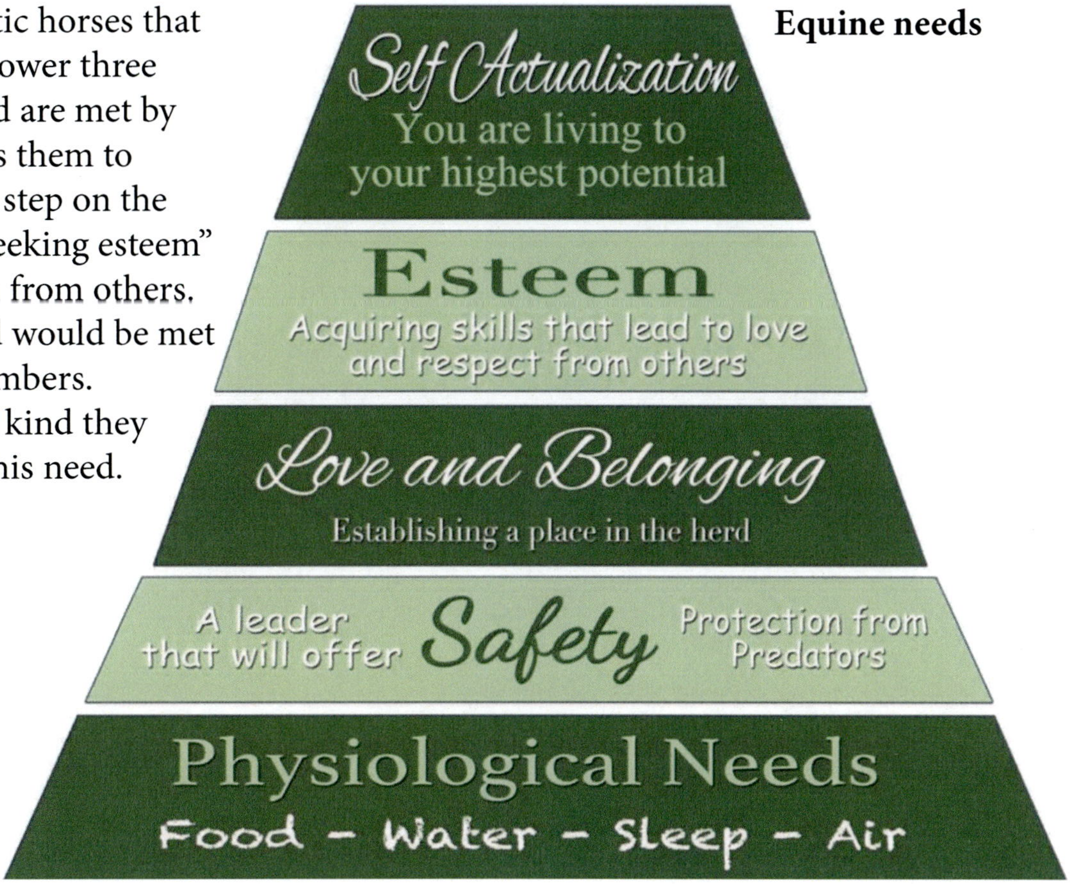

So what exactly does "self esteem" look like?
In human terms it equates to "likes on Facebook", acknowledgment for a job well done or accolades for your personal achievements. We crave recognition and appreciation from others to validate our self-worth. One of the most popular answers when you ask small children what they want to be when they grow up is "rich and famous" or rock star or movie star.

In equine terms horses strive to please us in return for our love and appreciation.
A pat on the neck or a "good boy" reassures them that we appreciate their efforts to do what we ask of them (and we ask a lot).
Man has devised numerous challenging equine disciplines which he trains his horse to excel in, ranging from show-jumping to harness racing. For the most part horses strive to figure out what we want from them and deliver to the very best of their ability. Despite the demands we place on them, many horses choose to form a strong partnership with their owner.

But! There is always a "but".
Some horses have figured out that when they are naughty they also get plenty of attention. Unfortunately for us, in the case of these horses their need is fulfilled with negative, not positive, attention (they don't differentiate between the two) making them hard to train.

Equine Survival Mechanism

Freeze, flight or fight

Experts agree that the natural response a horse will have when faced with danger is the fight or flight response. This is common among all prey animals and essential for their survival.

The experts also concur that his first response will be to flee.
The horse may feel threatened by a variety of perceived dangers both real and imaginary. In the wild his natural predators would be mountain lions, wolves, bears and maybe hunters. His heightened sense of self-preservation may cause him to bolt from loud noises such as gun shots, fireworks or even noisy vehicles.

If the horse is trapped with nowhere to run to, cornered and feeling threatened he may attack using all means available to him, all four hooves and teeth. Fortunately for the most part horses are gentle, placid animals, but make no mistake if a horse decides to attack a human the best course of action is to flee.

Fight or flight is only two/thirds of the horse's natural response to danger.

His first response is to **freeze.**
He will stop dead in his tracks become completely motionless. His eyes are wide open and nostrils flared (so he can take in extra air needed if he decides to flee).
In this moment he is making a decision - is the threat real or just something new?

All riders will have experienced this "freeze" reaction at some time or other.
Set the scene:
You are out for a trail ride when your horse hears a rustling in the trees. He spins around to face the direction of the noises and plants his feet firmly on the ground, stretches his head and neck out to see what the hidden threat might be. After a few moments he decides it was probably only a bird and resumes his trail ride.
Or perhaps - he spots a plastic bag lying on the ground - this piece of garbage was not there the last time he came down the trail; it's new and looks pretty scary.
Again he freezes to determine his next course of action. As riders we all know how scary plastic bags can be. The horse decides this threat is too great, as he spins on his heels and runs back to the safety of the barn.

It is in this moment that a trainer or rider can soothe and measure the horse and prevent the incident from escalating.

To test this theory we conducted a little experiment designed to show how horses freeze for a few seconds before they decide on their course of action in sight of danger or uncertainty. This is much like a deer freezes in the headlights of an on-coming car at night.

The objective is to see how long that "freeze" lasts before the horse decides on his next move.

The experiment :

* You will need an umbrella - and good running shoes.
* Carrying a folded umbrella, walk into a field of horses and call them to you (this works really well with youngsters).
* They will move in close to greet you.
* Turn away from them and encourage them to follow you - youngsters will think this is a good game.
*When they are all following you - pick up the pace and start running.
* Then spin around and pop open the umbrella in their direction.
Observe and time their reaction.

The inquisitive yearlings amble closer to check out this "new" person in their paddock.
Melissa crouched down, and used her body language to appear submissive and non-threatening.

She stood up, turned to face the horses and popped open the umbrella.

They jumped back - then froze for a few moments.

Look how tense and alert they are, compared to the first picture.

After only a few seconds the horses decided the umbrella was probably going to eat them so they had better run.

Notice the look of fear on this horse's face: eyes wide open, nostrils flared. Even though he is standing still he is leaning backwards away from the dangerous umbrella.

After the experiment, the horses were fed and gently groomed. Our goal was to demonstrate a theory, not leave the youngsters traumatized.

The **freeze** is not just a reaction common to animals - humans also react this way but for a slightly different reason.

They freeze to become invisible in the hope that whatever predator is posing the threat won't see them, similar to playing dead.

Set the scene:
You are 10 years old, sitting in a history class at school.
The teacher looks at the class and asks for a volunteer to come to the front of the room and read out loud from the text book. Reading out loud is your biggest fear, you stumble over the words, skip lines and then get lost in the chapter. You are sure the other kids are secretly laughing at you. Even the thought of standing at the front of the classroom, looking out at your classmates, fills you with dread.
Not a single kid in the class puts up their hand to volunteer - so the teacher announces, "OK I will pick someone to come up and read."

You immediately **freeze** - the goal is to become invisible so she doesn't pick you. If you are very still so as not to draw attention to yourself and make NO eye contact the teacher won't see you and hopefully she will call on some other poor kid.

For many people the fear of public speaking is worse than going to the dentist and having a root canal.

Herd Dynamics

Family feuds and favorites

I want you to take a moment and think about why horses live in herds.

We know horses are very social animals and form herds not only for protection, there is safety in numbers, but also to satisfy their need for companionship.

In the wild most horses in a herd would be blood relatives, so in reality they are just one big extended family. Just like human extended families there may be some bickering and disagreements from time to time. But just like human families blood is thicker than water and they get over their squabbles and quarrels.

Only in captivity do we force horses to live in unnatural and non-related herds.
Some owners don't even believe in a co-ed existence and separate the mares from the geldings.

The good news is horses are very resilient and comply to whatever domestic living arrangements their owners provide. However the natural desire to "belong" is extremely strong (third rung on the hierarchy of needs). This means when a group of non-related horses are kept together they will form a herd and eventually establish the same dynamics as a wild herd.

The most dominant and confident horse in the group will rise to the top becoming the alpha of the group and inevitably one horse, probably the youngest or most submissive member of the group, will become low guy on the totem pole. All the other horses will find their place somewhere in the middle.

Have you ever wondered how each horse knows "who is who" and who is higher in the herd hierarchy than whom?

A very easy and a simple experiment will reveal exactly who is Top Horse and also how they use their body language to communicate this information to each other.

This experiment works best when the horses are kept in a large but bare paddock and they receive their food at regular times throughout the day.

Conduct the test just before feeding time when the horses are both hungry and anticipating their meal to be delivered.

The experiment:

Walk into the middle of the paddock carrying just one slice of hay. It's important to go into the middle of the paddock as you don't want any of the horses to feel trapped - they need to have freedom to move at will. Call the horses to you and drop the hay on the ground and move a safe distance away from the hay, preferably leave the paddock.

One slice of hay is not enough for all the horses, which means only the alpha horse will get to eat. He or she will scowl and give threatening looks (which is usually all it takes) to the other horses, using his or her body language to loudly say "This is mine and I am not sharing." The lowest horse in the herd will quickly recognize there is not enough hay for everyone and will wander off knowing he is definitely not getting any. Some of the other horses may try to get near the hay but quickly get the message from the alpha horse.

Watch for a horse who seems particularly persistent and doesn't seem that bothered by the alpha horse's dirty looks. He circles around, trying to get close enough to steal a mouthful of hay. This horse is the "second in command" in the absence of the alpha horse he will quickly take his place.

Disclaimer: Of course, when working with horses be prepared to expect the unexpected.

Once the alpha horse had let everyone know "who was in charge" he settled down and allowed the other members of his herd to share the hay.

New Horse Introductions

Most horses are bought and sold several times throughout the course of their lives, which means they will have a variety of different homes and stable companions.

I don't know if, when a horse is sold and goes to live at a new barn, he misses his old stable buddies. I can't think of a way to prove or photograph this concept, so for now we will just have to speculate on any emotions he may be feeling.
What I know for sure, is he will have to be accepted by the horses at his new location before he will be allowed to find his place within their herd. Just because they are all horses doesn't mean they will automatically be friends and get along.

Human equivalent
The closest equivalent to this scenario is a child being taken from his parents and placed into a foster home where there are already three or four children. There will be a period of adjustment before the new child is accepted by the other kids and included in their games. During that adjustment period the new child is somewhat an outcast. Can you imagine how sad, alone and wretched a child must feel when forced into this situation?

I have witnessed many new horse introductions and they all play out in a very similar way.

Before the trailer containing the new horse arrives at the barn the existing herd is grazing peacefully in the pasture. When they hear the trailer arrive and sense new horses in the yard they (as a herd) move towards the gate.

The new horse or horses are excited to meet the existing herd and trot towards them.

For a few moments the horses all sniff each other and a few squeals and snorts are exchanged- there also may be some kicking and striking out - but this only lasts for a very short time. When the existing herd have checked out the new guys they usually all take off as a group leaving the newcomers alone. The new horses seem overwhelmed by the experience and walk slowly towards the group.

They keep a respectful distance between them - they have not been invited in yet. Occasionally small groups or individuals from the herd will come and check out the newcomers and give them a second look over. A little snorting may ensue and they quickly return to the comfort of their herd.

It is really quite sad to watch - the new horses just follow the herd (at a distance) until they are accepted - this might take days or weeks.

Existing herd, enjoying a peaceful afternoon.

New horses arrive at the yard.

Existing herd comes to check out the newcomers.

Initial introductions consist of sniffing and squealing.

The three new horses are chased away.

Existing herd gallops away as a group.

One horse looks back to make sure the new horses are not following too close.

New horses are shunned from the herd, but stick together for companionship.

Human equivalent

I want you to think back to the physical education class in grade school.
The teacher selects two of her most talented students to be the "Team Captains".
Each captain takes turns picking his team from the line-up of class mates.
Each child was standing tall desperately wanting to be wanted and picked early by the captain. As the teams got bigger the kids who were left felt a growing anxiety; everyone knew if you were picked last no one really wanted you on their team and you were considered a loser.
I am not sure if this practice still happens in schools, but it was a gut-wrenching and humiliating experience, especially for the kids who were not sporty or athletic.

Equine Body Language Visual Reference Guide

It is a popular theory that when animals communicate nonverbally, each sign has only one meaning, that they do not have the capacity to control and vary the meaning of their body language depending on the circumstance and that they are instinct-driven and have very limited reasoning or cognitive abilities.

I do not agree with that theory.

Anyone who has ever owned a horse will attest to the fact that they experience mood swings. Even the most placid broke gelding can have his "off" days, and chestnut mares did not earn their reputation for being feisty by accident.
Some horses are naturally confident and flamboyant in their gestures and movements and others are more subdued and quiet by nature.
Some horses exhibit a wide range of facial expressions and others seem pan-faced.
Despite the differences in communication styles they seem to communicate with each other extremely well.

It is also said that horses only react to immediate stimulus and are truly living in the moment.

I also disagree with this theory

Again, anyone who has ever ridden a horse knows they have excellent memories otherwise it would be impossible to train them. This means they recall what they learnt yesterday in order to be able to move forward with the next lesson. I also think horses are capable of worrying about the "what if". Perhaps this is not in the same way humans do, but you only have to watch a stallion with a group of mares to clearly see he is worried about predators, both real and imaginary.

The following pages take you on a visual journey where you will discover a wide range of equine body language. In order to fully understand the context in which the interactions were taking place, I have provided a little bit of the back story that led up to the equine conversation. The interpretation is based on past experience and research but of course, it is just an educated guess as no one knows for sure what horses are really thinking.

The human analogies help to draw a parallel scenario and empathy for what the horse may be feeling and trying to communicate. The goal is to look beyond the obvious and seek a deeper understanding which very often involves a little detective work that takes into account all the surrounding elements and individuals involved.

Disclaimer: I do not believe horses think the same way as humans nor do I believe they have the same understanding and values. Never the less after countless hours of observing horses and their natural interactions I am certain they are much smarter that we give them credit for.

Set the scene:
Two foals greet each other and one appears to be talking to the other. His lips are moving but no sound is coming out.

This action is called "snapping."

It is an appeasement gesture, which means "I am just a baby - Don't hurt me".

Foals will "snap" at other horses, people, trees and pretty much anything that is bigger than them.

He somehow knows how important it is for the other herd members to like and accept him. In the wild a rejected foal would not survive for very long.

Note: horses out-grow this behavior as their ego and confidence develops.

Likeable

Human Equivalent
Set the scene:
Humans were not designed to live solitary lives. We have a deep-rooted need to be loved and accepted. In fact the continuation of our species depends on it.

We want others to like us!
So we consciously modify our behavior to achieve that result.
Likeable people are:

Polite not Rude
Kind not Mean
Generous not Selfish
Friendly not Stand-offish
Fun and Outgoing not Miserable

You get the picture.
We silently display many of these attributes through our body language.

Note: Dale Carnegie's book - *How to win friends and influence people* is a best seller because deep down we all want to know that secret.

Everyone knows that when a horse pins his ears flat to his head and looks grumpy - you better watch out as he is obviously upset.

The trick to understanding equine body language is to try and see the bigger picture.

Sure the horse in the picture looks really angry - the question is WHY?

What situation or actions triggered this response?

Are the "dirty looks" directed at a human or another horse?
Remember this is how horses communicate, he may be telling another horse to: get out of his way or move faster. The look may mean don't come any closer. Horses all speak horse and understand each other perfectly. So pay attention to body language and responses of the other horses. Do they listen to his demands or is he chastised for rude behavior? This will depend on his place in the herd.

Note: Hopefully the angry look is not directed at you.

Angry

Human Equivalant

Set the scene.
Although we don't have the ability to wiggle our ears.
We do posess the ability to convey strong emotions with our facial expressions and body language.
The cause of this mans anger is not limited to living creatures.
He could be upset about:
His car, computer the weather or
even a play made by his favorite sports team.

Note: Just because a person has an angry outburst doesn't mean they are a bad person.

CAUTION

These pictures are quite extreme. The bay horse is a young, aggressive stallion.

All these images show how a horse will show frustration and irritation.

Tail Swishing
When annoyed a horse will tense his buttocks and swish his tail back and forth. This looks very different from when he is using his tail to swish flies away

Note: Horses give clear indications of how they are feeling and what they plan to do about it. Our job is to recognize those signs.

How to Recognize the Warning Signs

Flattened or Pinned Ears
There is no mistaking the look on this horse's face. His head is low and pushed forward, ears pinned flat to his head and an intense look in his eye.
He is mad at someone or something.

Charging Towards You
When a horse charges towards you, ears pinned, shaking his head from side to side - you better get out of the way.
This is not a friendly gesture. It is usually seen when a stallion is warning another male to stay away from his mares.

Teeth Bared
Biting is a show of dominance; the more dominant horse will bite a less dominant horse to show his authority and gain respect. It is normal equine behavior - but if humans get in the way the results can be extremely painful.

"*Horses are dangerous at both ends and tricky in the middle*" *Sherlock Holmes*

Set the scene:
One horse in the herd is always on his own.
He appears to be anti social.

He doesn't seem to have a friend or fit in with the other horses.

When hay is given to the herd he hangs back, sometimes until all the hay is gone.

It's quite sad, he looks so lonely.

This horse may never have been socialized as a foal (allowed to spend time with his mother and other horses in a pasture). Equine social skills are learned within the first year of a foal's life. If this crucial step is skipped, the horse honestly does not understand how to interact with other horses.

He wants to fit in and be accepted into the herd, but he just doesn't know how.

Anti-Social

Human Equivalent

Set the scene:
An only child is home schooled.
He is nurtured, loved and educated by his mother and father.

His mother takes him to the park to play, but the child looks sad and distant.

He doesn't know how to make friends. He doesn't even know how to talk to other children. All his social interactions have been with adults.

Socially Awkward

Note: A better description of this situation would be socially awkward and lacking the skills to successfully interact with peers.

The secret of really connecting with a horse is "simple" but not "easy".

You must learn to "quieten" your mind and spirit" and just "be".

This is not easy as we tend to over-think situations and have a million things racing around in our brain.

Note: Don't try to rush this process. It might not happen the first time you meet a new horse. The connection and trust may take a few attempts.

How to Connect with a Horse

Step 1 Go into a place where the horse is happy and relaxed - maybe his pen, field or stable.

Step 2 Empty your mind - push all you worries and to-do lists aside.

Step 3 Stand or sit for at least 5 minutes, don't DO anything.

You are just sharing his space

Step 4 Ignore the horse - don't make eye contact or talk to him - just be still.

Step 5 Focus on your breathing, just in and out. Don't try to change it.

After a while you, will become more aware of the sounds around you, birds singing, wind in the leaves even the sound of the horse breathing and eating.

You will know you have connected if the horse walks over to you.

That is OK, just stay calm and let him check you out.
Or if he looks at you (in an acknowledging way) and cocks one ear towards you.

For some people, to do nothing and quietly hang out with a horse is extremely difficult, but it's impossible to form a connection if you are distracted.

Set the scene;
The bucking horse is in the metal chute, this is not his first rodeo. He knows what is coming next. This horse fiercely objects to cowboys sitting on his back and digging their spurs into his sides.

All he can think about is putting as much distance between himself and these predators as possible.

At risk of hurting himself, he tries to climb out of the chute.

Note: Horses have only two responses to danger - fight or flight. This horse didn't escape the chute but he did give the cowboy one hell of a fight when he tried to ride him.

Anticipation

Horses remember when they are treated well and also who treats them badly.

The cowboy lowers hiself onto the hot, sweaty and visably distressed horse.

He knows this is not going to be pretty, in fact he knows he will very likely end up sitting in the dirt.

All he can think about is lasting 8 seconds and getting lifted off this time bomb of a horse by the pick up guys, before it explodes.

He knows there is a serious risk of injury, but he yells "open the gate".

Just like last time - the horse dumps the cowboy in the dirt

Note: Only humans willingly put themselves in danger. All other species try to avoid it.

Snapping

It's hard not to chuckle, when you catch a glimpse of a foal exhibiting this "snapping" behavior, the facial expressions range from cute to bizarre. Maybe that is why it works - the foal is trying his hardest to be cute. In this instance he was trying win over the mare who was flashing him a "warning look".

Horses communicate much more through postures, gestures and expressions than they do with their vocal cords, making them experts at de-coding the body language of other horses.

Did you know?
This skill is not limited just to equines.
Horses have the ability to interpret the body language (mood and intentions) of other species including humans.

Set the scene:

Out at the cattle ranch it's time for the calves to get imunization shots and ear tags.
The cowboy has to "sort" the older cows away from the calves and put them in a seperate pen.
He is riding a "cutting horse" who seems to love this type of work. The horse is able to keep the calf from returning to the herd with very little assistance from the rider. He has the ability to read each and every movement of the calf and anticipate where it will go. The horse appears to be dancing as he shifts weight from one front leg to the other in readiness to head the calf off should he run in the wrong direction.

Body Language

Nonverbal, usually unconscious, communication through the use of postures, gestures, facial expressions etc.

Human Equivalent

We all have a natural ability to read body language. Some people choose to boost their awareness to help them excell in their chosen field. This is especially evident in sports where anticipating your opponent's next move could be the secret to your victory. Ignoring this basic skill could prove painful if you are a Boxer or Martial Arts fighter. Less voilent sports also rely on competitors anticipating each other's next move, games like Tennis and Badminton.
Or the classic game of Poker, where it is crucial that your body language and facial expressions don't give away your next move.

Set the scene:
A young inexperienced stallion is turned into a paddock containing two brood mares (who are in season).

He is excited at the prospect of breeding the mares.

However he has no manners or social graces. His method of courtship is to bully the mares into submission.

This disrespectful behavior is not going to help him achieve his goal.

Note: In equine society, it is the mares who are responsible for teaching the foals how to behave in a courteous and respectful manner. It is important that foals are "liked" and the herd accepts them, so good manners are very important. The brood mares gave the stallion a "refresher course" on manners that afternoon. *Conclusion -He didn't get to breed either of the mares that day.*

Bad Manners

Human Equivalent

Set the scene:
You have been set up on a blind date.
On first blush the guy looks OK.
He is well dressed, tall and handsome and he even smells good (big improvement on past set-ups).
However he answers every call and chats animatedly to whoever is on the other end, turning his back and completely ignoring you. You give him a disapproving look but he doesn't seem to notice. Eventually you decide to gather your purse and leave.

Note: Hopefully he gets the message. But more likely he will think "how rude" she left without saying goodbye.

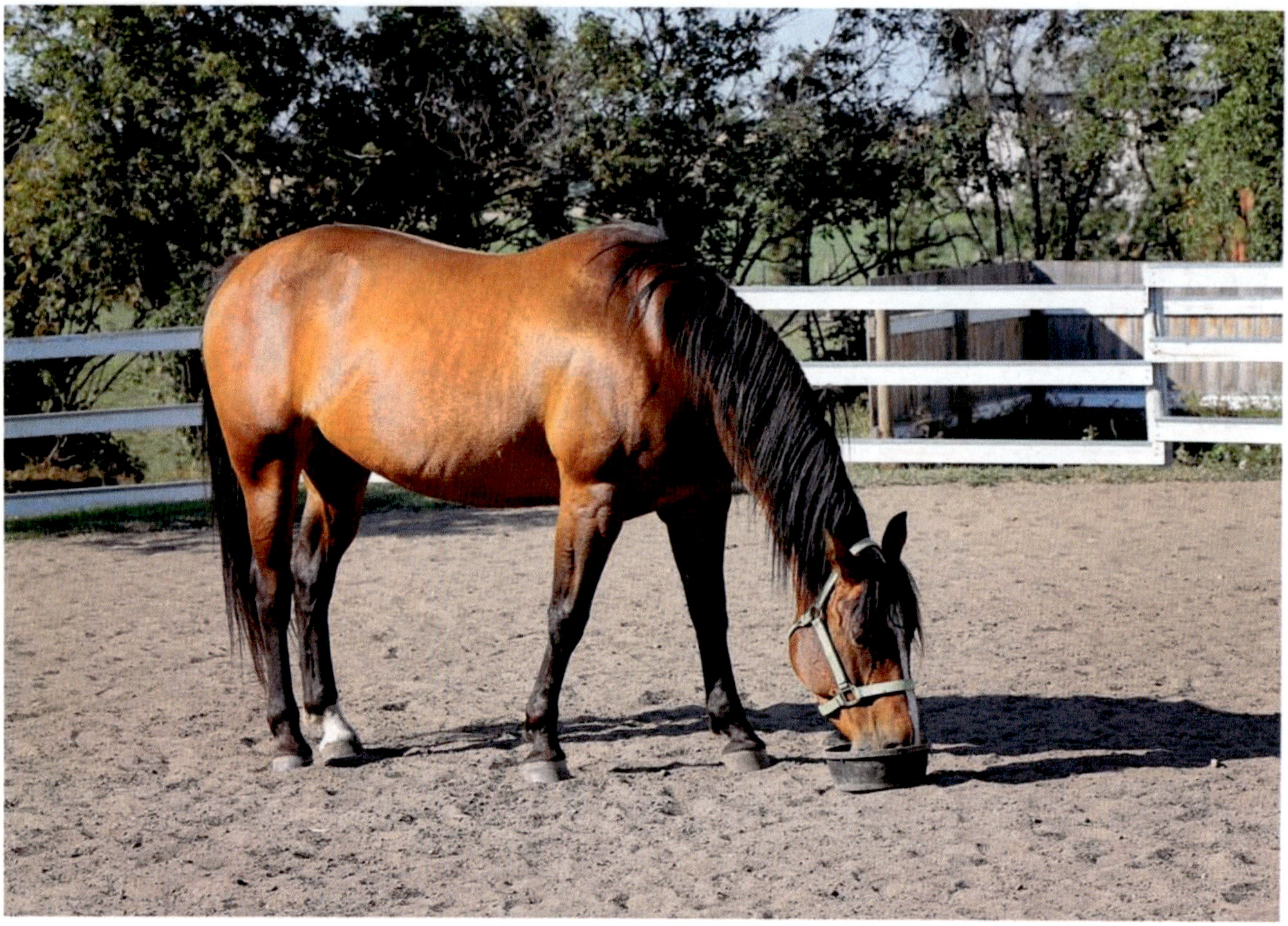

Bolting their Food

To devour the food as fast as possible without stopping.

Although this is more an example of a behavior than body language the action speaks volumes.

It is natural for horses to graze, eating little and often. Their digestive system is designed to continually produce enzymes that break down their fibrous diet in the gut. When horses are put on a restricted diet or have no access to food for long periods of time, they feel like they are starving. When they do eventually get food they eat it as fast as they can. Some horses get possessive over their food and aggressive to anyone, equine or human who comes near while he is bolting down his dinner.

This man-made problem has an easy solution - emulate nature and feed little and often.

Note:
Horses learn how to groom each other at a very early age.

Set the scene:
Once foals are comfortable leaving the safety of their mothers' side it is normal for them to seek the company of other horses their age.

All the youngsters hang out together, they play, graze and engage in social grooming. Scratching and gently nibbling each others' backs and withers. This physical touch enhances bonding and a feeling of belonging.

Social grooming is a very important part of equine social life.

Bonding

Human Equivalent

Set the scene:
For her birthday Sally asked her mother to host a slumber party.
The girls would create their own mini spa, paint each other's nails, braid each other's hair and give each other facials.

All-girl get togethers allow girls to be themselves, forge strong relationships and create unforgetable memories. Sharing secrets and popcorn helps form bonds that last a lifetime.

Set the scene:
It's a beautiful sunny day at the horse show. All the competitors in the western equitation class are lined up waiting for their turn to complete the pattern.

As you look down the line many of the horses look "bored."
Heads drooping, ears lopped out to the sides and eyes half closed.

But boredom is a human trait so the question is, do horses get bored?

By definition: Boredom occurs when you have little control over your situation.
Waiting rooms, lectures and airline gates are all places where you have little control.

This certainly describes the horses standing in the hot sun waiting for their turn.

Note: You may see horses in a pasture, standing around, heads drooping, eyes half closed, and ears lopped out to the sides.
The only difference is that those horses have a choice. They are free to move around at will, graze, snooze or even gallop around if they choose to.

Are they bored or just relaxing in the sun?

Bored?

Human Equivalent

Set the scene:
You spend the whole day lying around with nothing to do but relax and daydream.

Set the scene:
Six horses were living together in a domestic herd, one of which was Stumpy, the miniature horse.

The grey was a new horse.

When the new guy was turned out into the pasture with the other horses he received a most unfriendly greeting.

Stumpy viciously attacked the new horse chasing him around the field, trying to bite and kick him.

Despite his small size Stumpy is a bully. Throughout the attack the other horses, stood back, looking on, not wanting to draw Stumpy's attention to themselves.

Note: Contrary to popular belief, bullying is not usually done by the alpha horse. Horses will test each other all the time. Horses who are insecure in their standing in the herd may bully those under them just because they can.

Bullying

Human Equivalent
Set the scene:
We all know what bullying looks like.

Most people also know what it feels like. It is when someone has the power to intimidate, threaten and ultimately cause them physical harm.

Unfortunately bullying is not limited to the school yard. It continues throughout life. It may occur in the workplace and social gatherings.

The abuse can be physical or verbal and more recently cyber bullying has been added to the line-up. A bully picks someone they know is weaker than themselves so that they can intimidate, harass or humiliate them to do their bidding. It is a misuse of their power.

Note: When it comes to spotting a bully - looks can be deceiving - they can hide in plain sight.

Set the scene:
Horses truly embrace the concept that "it takes a village to raise a child".
All the herd members feel responsible for the care and protection of all the young, not just their own.
In the picture you will see 7 mares babysitting 3 foals.
The babies are tired and need to rest so all the mares put aside their personal desires and form a protective ring around the youngsters and stand guard whilst they take a nap.

Note: You might expect this kind of behavior from wild, not domestic, horses. Thankfully man has not erased compassion and co-operation from their nature.

Co-operation

Set the scene:
There has been a massive flood and many people are trapped in their homes.

Volunteers step up to help rescue the victims of this natural disaster.

Note: Modern living has somewhat eradicated our desire to work together for the good of all. Very often it takes a disaster to bring out the best in us.

Set the scene:
When horses get colic (tummy ache) they don't just suffer in silence, they use their body language to convey what is wrong.

He will look towards his stomach and gently bite at it.
He may kick upwards towards his stomach again pointing to where it hurts.

He may be listless and off his food.

If the pain is really bad he may break out in a sweat.

Note: We must assume they do this to let other horses know they are sick and in pain. As horse owners we can pick up on these outward signs and help, by either removing the cause or treating the symptom.

Tummy Ache or Colic

Human Equivalent

Set the scene:
When we experience a pain it is natural to put our hands over the place that hurts.

This body language is extremely useful and is evident from a very early age.
It helps to pin-point the root of the discomfort.

Just think about the last time you experienced a headache or backache.

What did you do with your hands?

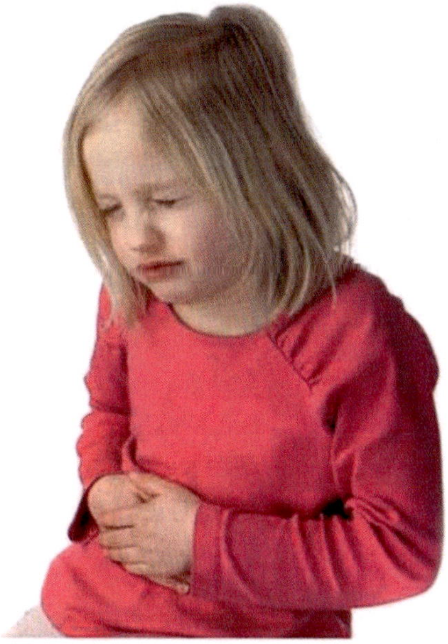

Note: Scientists found that gently rubbing a painful part of your body really does make it feel better. However rubbing away pain only seems to work when we do it to ourselves.

Set the scene:

The lady in the picture walked into a field to take pictures of the horses in their natural environment. Her intention was to capture shots of them "just being horses" and doing what horses do.

On this particular day they were extremely curious about a new person in their pasture. They insisted on checking out the camera and crowded around this interesting addition to their living space.

Because the lady was crouched down and motionless (a submissive posture) the horses did not feel threatened or intimidated by her presence.
In fact they invaded her personal space to satisfy their curiosity.

Curiosity

The desire to know or learn

Human Equivalent

Set the scene:

It is Valentine's Day and a beautiful bouquet of flowers along with a large box of candies are delivered to your door. The card attached says "Be my Valentine" but no signature. You are thrilled with the gift but are not romantically involved with anyone right now. This gift really is from an unknown admirer.
He obviously has good taste so you are extremely curious to discover who it is. You call your girl friends looking for clues, do they know who could have sent the Valentine's gift. It's driving you crazy you need to know who this mystery man is;
after all he may be Mr. Right.

Understanding Equine Body Language - "*Silent Conversations*"

Set the scene:
Do horses get depressed
and what would that look like?

Horses can and do feel depressed, this is may occur when their basic needs for equine companionship and exercise are not met.
A horse who is constantly stabled and denied access to other horses may show signs of depression.
It can manifest as weaving and box walking, both outward signs of stress and frustration.
Lack of interest in food causing weight loss.
He may hang his head low facing the wall at the back of his box sulking or he may be visibly grumpy exhibiting a change in his nature.
A horse kept in solitude has every reason to be depressed. Horses are social animals and crave the company of their own kind.

Note: In the wild horses live in tight knit family groups; everyone in the herd is related. Only in captivity are they forced to live with horses who are not family members and sometimes not even the same breed.

Depression

Human Equivalent

We understand the sadness horses must feel when they are depressed. We may even envy their simple cure.

Unfortunately depression in humans can be caused by a multitude of things and just being around other people is not necessarily *an instant fix.*

Very often people feel most alone when they are in a crowd.

Set the scene:

One horse biting the rump of another horse may appear to be mean, but it is quite normal for horses to challenge and test each other to establish their rank in the pecking order.

The more dominant horse bites a less dominant horse, usually for failing to move away in response to a lesser gesture such as ear-pinning, or a glare.

Note:
Biting is a form of communication. Youngsters will nip and bite in play, however if two horses decide to fight for dominance the biting will be more aggressive.
To interpret the message behind the action you to need to take into account how well the horses know each other, their relationship and the situation.

Dominance

Human Equivalent

I am not suggesting you wear a superman costume to appear strong and confident, however copying the "Superman stance" will.

The way you stand (your body language) is easily read by people around you.

The way you hold your body speaks volumes.

Set the scene:
It is mid-July and the temperatures are soaring.

One of the horses wades into the pond to cool off.

When horses have access to natural bodies of water they not only drink from it even if they are provided with fresh clean water they will go for a dip. Some will even roll around in the shallow muddy areas around the banks.

All horses have the ability to swim and will happily swim across a stream to reach better pasture but some seem to enjoy the water more than others.

Hydrotherapy:
Equine therapy pools for swimming horses are becoming increasingly popular as part of the training program for equine athletes, especially race horses.
As with people, swimming is beneficial in horses for conditioning the muscular and cardiovascular systems without subjecting the limbs to concussion.

Love of Water

Note: Perhaps horses already knew about the healing powers of water, and the soothing properties of cool streams on hot and swollen limbs.

There's something about water that draws and fascinates us. It is the most omnipresent substance on earth, along with air the primary ingredient for supporting life as we know it.

Set the scene:
The grey has just been introduced to an existing herd of horses *(mares and geldings).*
After spending a week in quarantine, he is desperate to find a friend.

He quietly approaches the paint, to see if he is friendly.

The paint is unsure of this "new horse" so he strikes out with his front leg establishing his personal space.

This is a "stay where you are - don't come any closer" gesture.

Message understood. The grey horse stood quietly allowing the paint to check him out.

Eventually the paint decided this new horse is OK and exchanged breaths with him.

Once the introductions were over the paint walked away inviting the grey to follow him.

Note: Observe how low and relaxed the head and neck are as both horses walk away.

Equine Introductions

Note: "Horses are dangerous on both ends and crafty in the middle." *Arthur Conan Doyle*

Familiarity

Reasonable knowledge of a person or place

Research survey:

Fifty horse owners were asked if their horses knew which stable was theirs and if given the opportunity to freely walk into any available stall, would they always go in "their usual" spot? The result was that 95% of the horses knew exactly where their stall was and made a bee-line for it, by-passing other open stable doors to get to where they were usually stabled.

The need for familiarity is very strong in both horses and humans. Familiarity makes us feel safe, comfortable and in control. Unfamiliar things and places make us feel uneasy so we strive to make them familiar by adding repetition and routine to the circumstance.

Human Equivalent

Remember your first day of high school, you arrived early, only to stand awkwardly waiting in the school yard. Eventually the bell rang and all the kids filed into the classroom. You look around for a familiar, friendly face, but don't recognize anyone from junior high. Then you spot a kid who looks just as bewildered and out-of-place as you and sit beside him. He seems OK; you exchange a few polite words. The next day is a little easier because once the bell rings you know where to go. After the first day, everyone will sit in exactly the same place for the entire semester.

Set the scene:
It is mid-summer and your mare is acting a little strange.
She is moody and disinterested in the work that she normally seems to enjoy.
However she is very interested in the male horses and will run up to them softly whinnying. Once she has their attention, she turns her hind quarters towards them cocking her tail to one side revealing her vulva.
She will open and close the lower part of her vulva which is known as "winking".

Note: When a mare is in season (ready to breed) her body language leaves nothing to the imagination. Her intention is to give the male a very clear message. There is nothing subtle or left to chance when it comes to equine flirting.

Flirting

Human Equivalent
Set the scene:
You work full time, volunteer at church play on a friendly sports team, cook, clean, shop and the list goes on.

Who has time for dating?

So you sign up with an on-line dating service. This is so much easier than trying to attract a suitable mate.

You create a profile that says "Look at me" and lets potential mates know you are available.

Note: When humans "flirt" it is a very complex and complicated dance, with lots of room for mixed messages. People often flirt just for amusement, rather than with serious intentions.

Set the scene:
You are visiting a friend's farm and would like a closer look at her horses. She is too busy to walk down the field with you, so you go alone.
You walk towards two horses who are peacefully grazing side by side.

They look up from the grass and start moving towards you, *even though they don't recognize you.*
Their heads and necks are in a low submissive position.
Their ears are forward and they have a soft kind expression in their eyes.

They are offering you a friendly greeting. If you mirror their body language and walk slowly and quietly towards them, your head slightly lowered, arms to your sides, the meet up will be pleasant.

The opposite is also true. For instance, if you run towards the horses, arms held high and outstretched, excitedly calling out to them *like a child chasing after an ice cream cart*, the horses will interpret those actions as aggressive and run away, putting as much distance between themselves and you as possible.

Friendly Greeting

Human Equivalent

Set the scene:
You have travelled to Thailand on holiday. The beaches are fabulous but you would also like to visit some of the temples.

You do not speak the language, so are unsure if you are allowed to look around.

A monk greets you at the door - puts his hands together, smiles and bows - then waves his hand towards the inside of the temple.

You acknowledge his invitation by mirroring his bow and quietly enter the building.

Note: There were no need for spoken words - body language was sufficient.

Set the scene:
The rearing stallion lives in a large pen.

The 6 horses at the gate are free to roam and graze in their 10 acre field, yet they choose to tease and interact with the stallion.

The frustrated stallion runs up and down the fence line; he even attempts to climb out of his pen.
He desperately wants to join the other horses but strong metal fences prevent him achieving his goal.

Pay close attention to the expression on the faces of the horses in the field, and the position of their ears. It seems as though each horse is sending a different message.

Note: One can only speculate as to their intention. Perhaps they are inviting him to join them in the field.

Frustration

Human Equivalent

Set the scene:
It's Christmas time and you want to be with your family so you are heading home for the holidays.
You arrive at the airport with plenty of time to spare.

You check your luggage and collect your boarding cards and everything is going according to plan.

Then, all of a sudden the display boards start flashing and your heart sinks. Your flight is delayed.

There is ice on the runway and the plane can't leave. You are stuck at the airport.

The most frustrating part is you dont know when the flight will actually leave or when you will get to be with your family.

Set the scene:
The stallion sniffs a pile of manure then lifts his head in the air and curls his lip up.

This posture is called " flehmen" which roughly translated means "testing". It helps him trap pheromone scents in the vomeronasal organs (VNOs). Pheromones are the chemical signals given off by other horses.
After he draws in the organic odor he curls up his lip to temporarily close the nasal passages and hold the particles inside.
Then an upward head tilt seems to help the airborne molecules linger in the VNOs, which are located under the floor of the horse's nasal cavity.

Stallions flehmen the most, but mares will also flehmen at the smell of birthing fluids.

Sex pheromones are definitely the most likely flehmen trigger, but they are not the only ones. Occasionally horses will react with an upperlip curl when they come in contact with an unusual nonorganic pungent odor.
Smoke from a fire, fresh paint, medicines or de-wormers are a few examples.

Note: Although the horse appears to be laughing, testing the air for pheromones is a serious matter.

Funky Smell

Human Equivalent

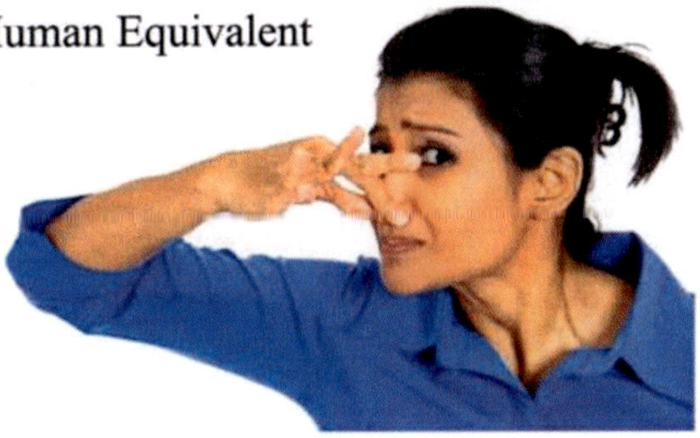

Set the scene:
There are lots of "smelly situations" in life and our "go to" response to funky smells is to close our nasal passages with our fingers to prevent the odor reaching our taste buds.

Note - This body language needs no explanation.

Set the scene:
These horses are part of a large mixed domestic herd.
The big chestnut is an old stallion (no longer used for breeding), however, he still feels the need to protect "his" mares from intruding males.

Today, the paint gelding is getting the brunt of his frustrations.

The small chestnut mare is just trying to stay out of the way.
She knows this show of aggression is not aimed towards her.

Note: Once the breeding season is over the horses will resume a more harmonious life together and the old stallion will feel less threatened by the younger geldings.

Human Equivalent
Set the scene:
You live in the country on a very private acreage.
It's beautiful but also a little secluded, which is why you need a guard dog.

This dog is wonderful with the family, but you see a different side to his nature when strangers come too close to the home (territory) he is protecting.

Go Away

There is a knock at your door, but you are not expecting visitors. So you grab your dog and open the door just a little. It's some guy canvassing for a political party so you really don't want to talk to him. In fact you don't even want him on your doorstep, so you allow the dog a little more leash. The guy backs away and leaves.

There is no mistaking the dog's message.

Note: You praise and reward the dog for protecting you - because his actions mirror what you are feeling. He has no inhabitions about standing on the porch screaming "Go Away."

Set the scene:
It is "move-in day" at the rodeo grounds. Cowboys and girls are setting up camp for the long weekend competition. The stock contractor who supplies the bulls and bucking horses unloads his animals and puts them in pens to relax before the rodeo starts.

This particular stock contractor, did not separate the horses and big ornery bulls. He put them in the same pen.

In a spirit of cooperation they ate together from the same pile of hay. I am not sure they were friends but they certainly were not enemies.

Go Along to Get Along

Spirit of cooperation

This well-known saying is an idiom that means to acquiesce.
"Go along to get along" does not necessarily imply agreement, just no active disagreement.

Humans who exhibit this trait may appear luke warm with no passion to follow their own goals or dreams. They just tag along for the ride.

Human Equivalent:

Some people get under your skin and push your buttons to the point where you feel like you will explode. It's impossible to get along with everyone!

Or Not!

Set the scene:

It is beautiful spring morning. The sun is shining and the birds are singing.

The horses seem "happy to be alive" as they canter animatedly across the field.

The foals show their exuberance by carrying their tails straight up (flagged), galloping and bucking just for the fun of it.

Very often the foals will engage in a playful game of catch me if you can.

Note: Although seemingly "happy to be alive" or playful behavior is more evident in the younger horses and stallions, older horses will occasionally run and buck just for fun when they are first turned out to pasture.

Happy

Human Equivalent

Humans have a variety of "nonverbal" ways to express happiness.

Think about the last time you did your Happy Dance.

What made you feel happy to be alive?

Set the scene:
It is feeding time in the barn and one horse is persistently banging the stable door with his front hoof.

It is noisy and very irritating.

He is desperately trying to get your attention and let you know he is hungry and wants his food NOW.

Note: It is natural for horses to move around and graze pretty much all the time.
When stabled, they rely on humans to provide their basic needs.
Not knowing when food or water will be delivered can cause the horse to become anxious and stressed.

Impatient

Human Equivalent

Set the scene:
It's Christmas Eve. You stand in line for 20 minutes to pay for your groceries, but the cashier makes a mistake, so she sends you over to the customer service desk for a refund.

The young girl behind the desk is chatting and giggling with her friend on the phone. She is completely ignoring you - she doesn't even acknowledge you are there.

You are trying to stay calm, but are in a hurry, so you loudly drum your fingernails on the desk.

You know it's rude but so is her behavior and you want to get her attention.

Independent

Wild horses are not dependant on humans for their survival

Although wild horses may not regard humans as natural predators, in the same way they view cougars and wolves, they maintain their independence by keeping a good distance between us and them.

Humans fulfill none of their needs so they have no desire to form a relationship with us.

Wild horses living in a herd depend on each other for their survival and success.

This is a typical stance when the wild horse spots a human. He will "**freeze**" for a second - head held high to get a good look at the possible danger, nostrils flared to take in extra oxygen in case he needs to run, ears pricked forward so he can detect any suspicious noises. He may stay like that as you approach carefully watching your movements. As soon as you get too close (he decides on what that distance is) he will turn and move away from you. The speed at which he takes "**flight**" is dependent on how threatened he feels by your presence. If trapped or cornered he would "**fight**" using teeth and hooves for his freedom.

Set the scene:
A large ball has been placed in the pen with a mare and her foal.

The mare completely ignores the ball but the foal is fascinated with this new object in his environment.

It may be something good to eat. Foals are extremely inquisitive and just like human toddlers will put everything in their mouths. This sometimes can be a problem.

Although the foal's antics may look like he is playing, remember he is a horse. First he has to make friends (establish if the ball is going to hurt him). Then he needs to determine who is higher in the pecking order (him or the ball). Finally he invites the ball to be part of his herd - now the ball belongs to him.

Note: What a foal learns in his first year will stay with him for life. It is important that he spends time with other horses and also youngsters his own age so he is socialized and learns how to play well with others. Weanlings play among themselves by play-fighting, galloping, bucking, and running with bursts of speed.

Inquisitive

Human Equivalent

Set the scene:
It's a rainy Sunday afternoon, there is nothing good on TV and the kids are looking for something fun to do.

"Let's play a board game."

Classic games:
These make you think and ask questions, try different options and strategies. They couple skill with a little luck and have a different outcome every time.

Cluedo was invented in 1944

Note: Cluedo taps into our inquisitive nature which is why it is so much fun.

Set the scene:
The foal is only 2 weeks old,
he is very shy and when other horses or humans approach he will hide behind Mom.

Mom is normally a very friendly mare, but she acknowledges that her baby is nervous and feeling a little insecure. She quietly helps the colt stay hidden by moving her body in front of him when other horses or humans approach.

The mare knows they are not a threat to her baby and her facial expression and body language are non-aggressive.

She is just helping her baby feel more comfortable.

Note: As the foal gains confidence and starts to graze a little farther from his mother, the mare will be less protective.

Insecure

Human Equivalent

Set the scene:
You have to give a presentation to the Board of Directors at work. The speech is well written, your family seem to think it's good,
but nothing seems to quell that nagging voice saying - "You can't do it, you will screw up - you are not good enough."

Insecurity is something that effects everyone at some point in their life.

Some people run and hide and others put on their game face and just do it.

Note: Psychologists attribute insecurity to a critical inner voice. Understanding where that comes from and challenging it is the first step to overcoming your insecurities.

Everyone know that horses use their tails to swish flies away !

But did you know horses use their tails as a form of expression as well?

Usually when a horse swishes its tail angrily back and forth it is because it is irritated by someone or something. The horse will also glare at the source of his irritation.
This is a warning and if ignored more aggressive action will be taken.

If the horse clamps his tail to his rump, he is frightened and may be about to kick.

A relaxed horse will carry his tail slightly lifted when moving.

If he is happy and exuberant he may carry his tail high up over the back or straight up like a flag.

Note: Foals very often carry their tails straight up when they are playing .

Expression

Human Equivalent

Babies have limited communication skills so you have to engage in a little bit of detective work (look at the bigger picture) to figure out what is the cause of their anxiety when they cry.

Perhaps they are:
scared, hungry, too hot or too cold
need a clean diaper
or just a hug.

You have to consider everything in their environment to discover the problem.

Set the scene:
This palamino stallion is living in a paddock with a couple of mares.

The girl is trying to catch him.

He has no intention of complying to her wishes.

He stands his ground and as she approaches he rears up.

He is asserting his dominance over her by being up in her face,
clearly showing a lack of respect.

Note: Respect is a natural part of a equine behavior and relates to the heirachy in the herd. Every horse in the herd needs to know exactly where they stand in the pecking order, otherwise fights would break out all the time. Heirachy is determined by dominant or submissive behavior.
Rearing is more of a stallion trait. Mares tend to exert their dominance with their back hooves.

Lack of Respect

Human Equivalent

Set the scene:
Mother and daughter are having a heated discussion regarding the daughter's curfew.

The daughter thinks there is no need for a curfew at all and wants to come home when she feels like it. Her mother insists they set a time that they can both feel good about. Mid-sentence the daughter raises her hand and turns her head away in a disrespectful "talk-to-the-hand" gesture.
She then storms out of the room leaving her mother angry and frustrated,
the argument is far from over.

Note: There is nothing new about the conflict a mother feels as her daughter grows up and demands more freedom. She knows this is a normal part of growing up, but part of her wants to protect her baby.

Set the scene:

Loner

Someone who prefers to be alone, and deliberately avoids the company of others.

This bay gelding is one of 30 horses who live together in a beautiful 40 acre pasture surrounded by trees. While most of the other horses graze together as a herd, there is a small group of 6 horses who seem to be special friends and hang out as a group. There is one horse who seems to prefer his own company and keeps himself very much to himself. Although this behavior occasionally manifests and the horse seems perfectly happy to be alone - it is not normal equine behavior.

Note: In the wild this behavior would be like committing suicide as the herd offers not only companionship but protection from predators.

Human Equivalent

There are many reasons why humans may choose a solitary existence in a very social world. Choosing to be alone may be a "coping mechanism" and the result of negative experiences. The loner chooses to remove him or herself from the negative stimuli so they can feel comfortable and at rest.

Set the scene:
The stallion senses danger and feels the need to move his herd to a safer place.

He lowers his head and neck
and swings it from side to side
in a snaking motion.

In this fashion he herds his family to safety.
The herd gets his message loud and clear and are compliant.

Note: The stallion moves his herd from behind - however the alpha female will lead from the front, inviting the rest of the herd to follow her to new grazing or fresh water.

Move

Human Equivalent

Set the scene:
You notice a duck and her babies
trying to cross the road.
They are in danger
from being squished by traffic.

Your instinct tells you that they need help.

You approach from behind the group,
bend over with arms outstretched
and gently shoo them to the other side.

Note - By your body language it is obvous to the drivers what you are trying to achieve.
In acknowledgement of that they stop their cars and wait for the ducks to reach safety.

Set the scene:
Two horses live in adjoining pens. They strain to reach each other through spaces in the heavy fence that separates them.
They crave the companionship and touch of another horse.

They stand breathing into each others nostrils for several minutes in a very intimate greeting.

One can only assume they are communicating in some way.

Note: When horses first meet they also touch noses and exchange breaths, however that first encounter is usually followed by squeals and foot stomping. As each horse realizes "I don't know you " which means they rely on **sight and smell** to recognize each other. When horses know each other, this interaction is quiet and relaxed.

Neighbors

Human Equivalent

Set the scene:
Two neighbors are gardening. One beckons to the other to come over to the fence.

They chat for a while before going back to their garden chores.

Note: The neighbors only need a casual wave to initiate the conversation because they know each other. It would have been different if a passer-by had beckoned from the street for one of them to come over.

Horses enjoy the close intimacy and friendship of other horses.
They will nudge and nuzzle each other as a sign of affection.

They crave that feeling of belonging that is present in all herd animals.

Some horses use the same body language to show affection to their human owners. It is better to return the affectionate gestures by scratching the base of their necks than feeding treats.

Nudge and Nuzzle

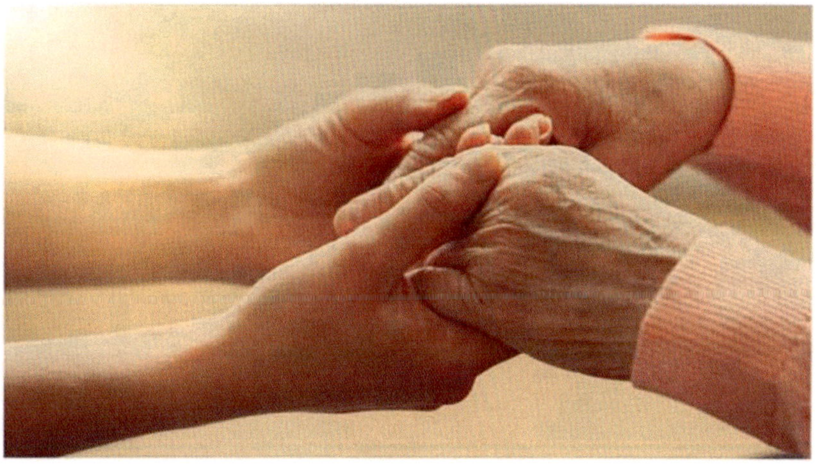

Human Equivalent

Holding Hands: Children hold their mother's hand, friends hold hands, lovers hold hands.
We hold the hand of someone who needs comforting, or to give courage in the face of a difficult situation. It is a simple gesture that humans across globe engage in, that allows us to connect in an affectionate (non-sexual) way.
It is impossible to feel "alone" when you are holding the hand of another person.

Set the scene:
It's late afternoon and the foal is lying down for a nap.

His mother would also like to take a nap but her first priority is to watch over and protect her baby.

So she stands over him as he sleeps keeping a watchful eye open for any signs of danger.

The name given to this behavior is "recumbency response".

Note: The mare may graze in a tight circle around her foal as he sleeps, always keeping her face turned towards her baby so she can see him.
As her foal grows the mare will become more relaxed, allowing him more freedom.

At around 3 months the foal will crave and seek out the company of other youngsters.
Together they will frolic and play and establish their place in the herd.

It's Natural to Nurture

Human Equivalent

Set the scene:
A new mother carries her baby with her at all times, reluctant to put him down.

When he is straped to her body she can continually watch over and protect him.

Her movements are greatly restricted and she has to modify her day-to-day actions to accommodate her new priority.

Note - A modern trend is to over-protect children and control every aspect of their lives "helicopter parents". It's an attempt to save them from any kind of accidents or discomfort. They deny the child the opportunity to learn from his mistakes.

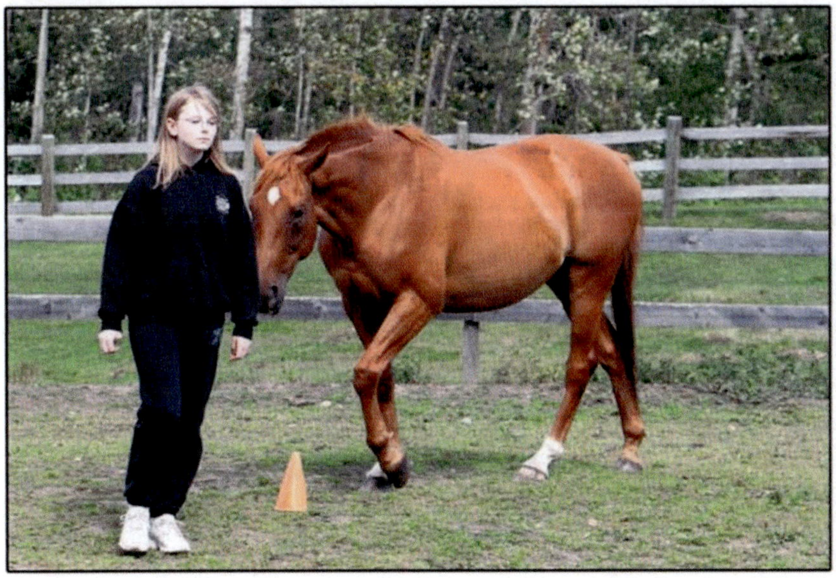

Set the scene:
Just as we try to understand and interpret the horses' body language, they too are trying to figure out ours.

In order to communicate we need to "think like a horse" and modify our body language and actions.

In this picture the girl is slowly walking in front of the horse. In horse language this is an invitation to follow.

Note: The horse is accepting the girl as his leader and placing her higher up in the herd hierarchy than himself. If she turns the horse will turn. If she stops the horse will stop. This little experiment would not work if the horse did not respect the girl or feel safe in her presence.

Observant

Human Equivalent

Horses don't understand human language. However they pay close attention to our body language so it is important we give clear positive messages.

Human babies can understand nonverbal communication before they develop the skills required to verbally communicate themselves.

This is why experts in child development are encouraging mothers to teach sign language to their infants, so they can use body language to have silent conversations.

Note: Humans put more value on verbal communication that body language which is why it is taught in schools. All other species communicate clearly using body language.

Set the scene:

Mountain trails that are kept clear during the winter months are first treated with salt spread by graders. Later as the snowfall increases, snow ploughs will go through the trails, pushing excess snow to the side of the road. This provides a free source of salt for the wild horses who are smart enough to take advantage of this valuable mineral.

Point to ponder:

How did the wild horses discover this free resource and how did they tell their friends?

Opportunist

Opportunism is exploiting circumstances to gain immediate advantage

Human Equivalent

For centuries coastal people have gathered washed-up seaweed from the beaches. Rich in nitrogen and potassium seaweed makes a wonderful fertilizer. As technology advanced the seaweed was dried and milled to making it easier to transport and use.

Set the scene:
Two wild stallions rear up to make themselves as big and intimidating as possible. They fight with mouths open, teeth bared using everything they have to win the fight.

Horses always, always have a reason for their actions. When they fight it is usually to protect what is theirs! This may be territory, food, or in the case of the stallion his mares and youngsters.

Young bachelor studs try to steal mares from existing herds so they can breed to start their own family group, but the dominant stallion does not let that happen without a fight.

Note: Look closely at the horse on the left and you will see an old scar on his hind quarter. This was not his first fight.

Ownership

Human Equivalent

Set the scene:
It's Halloween and the kids are out in their costumes "Trick or Treating".

They come home with a huge bucket of candy which they tip out onto the kitchen table to sort through.

Dad comes in and wants to steal the "good" ones - the Caramilk and KitKat bars.

So the kids try to cover their candy with their arms and swat at Dad's attempts to steal their candy.

Note: This example may not be as violent as the battling stallions, however the "feeling" is the same. Dad understands the children's body language and their frustration at trying to protect what is theirs, but enjoys teasing and harassing them anyway.

Horses have no problem when it comes to “finding their passion”.
The intense desire to reproduce, create a family and ultimately a herd is the third rung on their “hierarchy of needs”. During mating season the urge can become so strong that some stallions put the need to breed ahead of eating. Their passion is so intense it is all they can think about. Fortunately breeding season is fairly short. Nature designed this pattern to ensure foals are born during the warm summer months when food is plentiful. Once all the mares in a herd are bred and pregnant the stallion can relax and resume a calmer way of life.

Passion

Intense emotion, compelling desire

Human Equivalent

We tend to associate finding your passion with work or play - discovering that one thing that really turns your crank. It could be music, sports, art or maybe you are passionate about sharing your knowledge and teaching others. If you could only do one thing - this would be it.
Only humans struggle to find meaning in life by searching for things and activities that produce such a compelling emotion in them. Finding their passion makes them feel truly alive.

Set the scene:
Horses all have a "personal space". Some demand a larger space than others but when they feel their space has been violated the message to back off is delivered swiftly.

Sometimes it may be a warning, ears pinned and swishing of the tail.

Other times they will kick out with both barrels.

Note: Close contact is reserved for those horses they know and trust.

Personal space

Human Equivalent

Set the scene:
You are in the line up at the grocery store. The guy behind you keeps inching closer and closer.

You don't think he is threat but he is making you very uncomfortable.

He is invading your personal space and you would really like him to back off.

You swing around to give him a dirty look. He doesn't seem to notice and keeps inching closer. Because you don't want to make a scene. You try to move away from him to give yourself more space.

Note: Sometimes we put ourselves in situations where there is no room for personal space. Places where people are all crowded together, waiting for a train at Grand Central Station springs to mind.
Knowing it's going to be crowded doesn't make it OK. It feels uncomfortable when your personal space is violated.

Set the scene:
This mare and her brand new foal are part of a large domestic herd. They live together in 40 acre pasture.

The mare is very protective of her new baby and when anyone (equine or human) approaches she moves around her baby to shield him with her body.

She will pin her ears back and give you "the look" you know the one that screams "leave us alone."

There is always a reason behind why horses do the things they do.

Note: In those first few days of her foal's life the mare will be extremely protective.
She must prevent her baby from imprinting on another horse. It is essential to his survival that he can recognize his mother by sight, sound, touch and smell in order to suckle when hungry.
If he tried to suckle from a different mare he would be rejected and go hungry or worse be kicked and hurt.

Protective

Human Equivalent

Set the scene:
A father carries his son on his shoulders through a busy street,

keeping him safe and away from strangers.

This is such a familiar sight that we don't even realize the powerful message behind their body language.

No words are spoken yet you know.
The father cares for his son and is actively protecting him from danger.
The son feels safe and loved and comforted by the strength and power of his father.

Note: Humans are predisposed to care for and protect their young, for longer than any other species.

Horses create 6-8 piles of poop per day and they seem fascinated by it.
You will often see horses sniffing piles of poop.

So why do they do that?

To humans horse poop smells like horse poop. Our limited sense of smell cannot detect a difference between one pile and the next. Manure is evidently a veritable treasure trove of information to horses.

They can identify which horse was recently there. Each horse deposits their own unique scent.

A foal may use this method to locate his mother.

A stallion uses this method to identify all the members of his herd and spot intruders.
He can also tell which mares are ready to breed by sniffing their poop and urine.

Recognition

Note: Horses obviously rely heavily on their sense of smell rather than sight to recognize other horses and people.

Human Equivalent

Set the scene:
Adult and toddler are playing a game of "Peek-A-Boo".

Although this is a fun game it also helps the child with face and voice recognition.

The child giggles with glee when a familiar face pops out and says "Peek-A-Boo".

Note - Although humans all have their own personal scent, our sense of smell is not sophisticated enough to identify individuals by their scent. We rely heavily on sight and sound.
We can recognize someone from a photograph (no sound) and we also recognize people by their voice alone.
When a friend calls we know who we are talking to by the sound of their voice.

Set the scene:
The horses are enjoying a sunny afternoon out in the pasture.

One horse is lying flat out fast asleep,
one is lightly dozing,
one is grazing and the other horse is looking out for danger.

As prey animals it is important that at least one animal in the herd stays alert and on the lookout for predators.

Note: Even though horses are able to snooze standing up. They will rest and snooze throughout the day, however they still need to lie down and experience deep REM sleep for at least 15 - 20 minutes per day. Foals, just like human babies, need more sleep.

Rest and Relaxation

Human Equivalent

Set the scene:
You are working the night shift and have been awake over 24 hours.

Your eyes are heavy and red,
you are moving very slowly,
you repeatedly yawn and stretch.
You are irritable and short tempered.

From your "body language"
it is obvious you are suffering
from sleep deprivation.

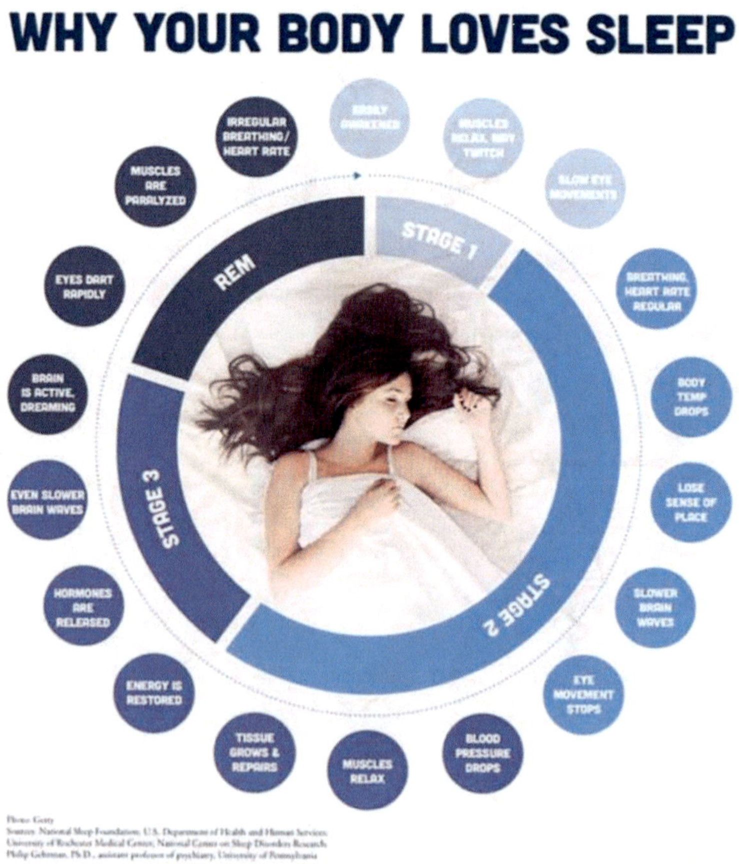

Note: We live in the "burn out" generation. We push themselves to the limit both at work and at home in order to have it all at the expense of essential rest and relaxation.
Only humans willingly and knowingly deprive themselves of sleep.

Set the scene:
It is early summer and the stallion lives alone his secure paddock. The mares live in their own large pasture. The stallion can see them but there is another pen in between that prevents the horses from touching.

The stallion sniffs the air. He can smell that some of the mares are in season and ready to breed.

He spends his day running the fence line calling out to the mares, desperately trying to get their attention.

He animatedly trots back and forth, head held high proudly displaying how big and macho he is.

This male is showing off.

Note: The urge for the stallion to breed is intense, fueled with testosterone. He will run the fence line for hours. During this time he will lose weight as he barely stops to eat. If the stallion is not going to breed with the mares it is kinder to keep him in a place where he can neither see or smell them.

Manly Display

Human Equivalent

Set the scene:
The cowboy gently lowers himself onto the back of the bull in the chute.
The bull grunts and lunges at the gate.

With one hand the cowboy grabs hold of the strap with his heart pounding. The adrenaline is racing through his veins, because he knows as soon as the gate opens it's just him and the angry bull.

Cowboys see bull riding as the ultimate challenge. Staying with the bull for just 8 seconds is a test to their courage, ability and manhood. The boys ride steers.

Note: A wide variety of competative sports illustrate the need for males to prove their strength and masculinity Body building, strong man competitions, wrestling and boxing, even rugby and football are to some degree for the same purpose. The display is more about impressing other men than impressing or attracting females.

Set the scene:

Many of the horses featured throughout this book form part of a large domestic herd who enjoy a carefree life in a large 40 acre parcel of land which is surrounded by a variety of trees and wild bushes. On first blush this pasture looks rough; there are weeds and thistles and the horses drink from a dugout filled with rain water.

These horses have access to a huge variety of grasses, shrubs and vegetation, unlike the manicured pastures that are enclosed by wooden fences and kept weed free.

Wild horses seem to have an innate understanding of how to self-medicate using natural herbs and vegetation. Eating just what they need to maintain good health, they do not "over-medicate". Domestic horses also have this ability when given free access to nature's healing herbs.

The horse in the picture is not being bad or destructive he is taking what he needs from the bark to feel better. The inner bark of poplars contains the compound related to aspirin called salicin. It can be used for fevers, as a pain reliever, and to reduce the swelling of injuries.

Self Medicate

Human Equivalent

Self-medication can quickly develop into addiction.
Initially the self-medication is to relieve either emotional or physical pain, whether using pills, alcohol or street drugs. Humans do not seem to have the same "self control" or understanding of when enough is enough. This new state of being is often preferable to dealing with the pain or resolving the emotional problem.

Long before pure chemicals were manufactured in labs, people used plants and herbs for medicine.

Self Preservation

A universal behavior that ensures the survival of an organism

Set the scene:

Travelling through the mountains, you spot a band of wild horses. Quietly you get out of the car to take photographs. As you move closer to get a better shot, the horses sense your presence. They recognize you as human. They have seen humans before and know you are not a carnivorous predator, however humans can be unpredictable, so best not to get too close.
They head into the safety of the bush putting more distance between you and them.

Human Equivalent

Walking home from work, you spot a homeless person sitting on the sidewalk. This seems odd as you have never seen this person there before. You feel uneasy. Thoughts, start running through your head.
Is he crazy maybe drunk or on drugs? Will he mug me for my money? There is no way of knowing if any of these fears are true. To avoid any trouble you cross the street and walk quickly away in the opposite direction.

Set the scene:
These two stallions live in adjoining pastures and for most of the year and they get along really well.
It's spring-time and 3 brood mares have been turned out into one of the pastures. This means one stallion has a job to do. He must protect his mares from any possible danger and other stallions.

The lone stallion can see and smell the mares and he too wants the chance to breed with them. So he runs the fence line calling out to the mares and showing off.

Such behavior cannot go unpunished so the fight is on.

In a testosterone driven display of aggression they spend much of their time bickering over the fence.

Selfish desire

Note: Their behavior is authentic. Neither stallion is trying to disguise his feelings or intentions.

Human Equivalent

Set the scene:
Two friends travel to New York to buy a wedding dress at a famous boutique sale. All the dresses are marked down to only $99 each.

The friends plan to help each other find the dress of their dreams.

Racks and racks of fabulous gowns are on offer. Unfortunately both girls spot the same dress at the same time and each girl desperately wants it.

Caught up in the frenzy of the sale, their friendship is set aside as they fight over the same dress.

Note: We try to hide our selfish nature because it isn't pretty.

Set the scene:
You go to the pasture to catch your horse because you would like to ride.

Your horse has a different opinion.

He turns his hind quarters towards you and looks away from you avoiding eye contact.

The action is deliberate. If you try to move into his range of vision he will re-position himself, making sure his hind quarters are still in your face.

Note - He obviously does not want to be caught and depending on how strong this feeling is, will determine his next actions. He may decide to run away or lash out with his back feet if he is truly determined to stay in the pasture with his buddies. Or his objection to being caught may be mild and he allows you to come around to his head and put his halter on.

Shun or Ignore

Human Equivalent

Set the scene:
You have had a fight (difference of opinion) with your spouse.

The fight is over but you are not ready to kiss and make up.

So when approached you fold your arms and turn your back towards him.
You avoid eye contact and
give him the silent treatment.

Note - Your spouse knows by your body language that you are not ready to forgive and forget just yet.

Set the scene:
The gate to the adjoining pen has accidentally swung open. The pony knows that this is not "his" pen and enters with his head down; he hesitates for a moment to make sure the pen is not occupied by another horse.

The coast is clear, so he slowly walks in, again keeping his head down so as not to draw attention to himself, or give away his intentions.

He had seen the human leave a bucket in the pen before heading out towards the pasture carrying a halter.

If he is quick, he can eat his neighbor's breakfast and be back in his own pen before he gets back.

Sneaky *Something done quietly behind a friend's back - so not to get caught*

Set the scene:
Two horses are really, really close friends. They graze together, move around the pasture together, sleep together and they seem inseparable.

The term for equine BFF's is "**Pair-bonded**".
Pair-bonding is an important and significant aspect of herd life. Horses choose a special friend typically, but not necessarily, of similar age, sex and size.
If permitted the friendship will last for life.

Pair-bonds allow each other into their personal space to engage in mutually beneficial behavior such as grooming and play. The primary biological and evolutionary purpose of pair-bonding is protection.

Note: Horses are social animals and crave the company of their own kind.

Special Friend

Human Equivalent

Set the scene:
During a ceremony rings are exchanged and worn on the third finger of the left hand,
silently declaring a bond
through sickness and health,
better or worse,
richer or poorer,
until death do us part.

Note: Love and belonging is a basic human need - the survival of our race depends on it.

Set the scene:

The horse was stabled overnight and in the morning his owner sets him loose in the outdoor arena to stretch his legs and get some fresh air.

There is a nip in the air but the horse seems happy to be outside. With a squeal of delight and a couple of bucks he sets off at a full gallop around the arena. Then he notices a pole lying on the ground and just for fun, he canters around a couple of times to jump over it.

Spirited

Displays animation, vigor or liveliness

Human Equivalent

The need to move freely is very strong in a lot of people.

Millions of people across the country get up early, lace up their running shoes and head outside to run just for the sheer pleasure of it. They may start running to lose weight or get fit - but they stick with the sport long after they have achieved their goals.

Many push themselves to go faster and farther.

Set the scene:
The horse is trotting along in his pasture when he suddenly notices something lurking in the grass.

He jerks to a halt, lifts his head high and pricks his ears in the direction of the mysterious object.

He raises himself up almost like he is standing on tippy toes.

He may also let out a snort.

Note: The horse has yet to decided on his next course of action.
He may decide the rustling in the grass was nothing to be worried about. In that case he will relax and carry on with his day.
Or he may believe there is something that will eat him lurking in the tall grass and take flight.

Human Equivalent

Set the scene:
You are turning the handle of a Jack-in-the-Box.

Even though you know what is going to happen when the lid pops open, the sudden movement when the puppet pops up startles you and makes you jump.

However once it is done you quickly calm down and carry on as normal.

Startled

Set the scene:
The horses are stabled overnight and let out in the morning.
The first thing they do is find a muddy area and roll in it.

They take great delight in rolling back and forth in the mud.
When they get up they give themselves a good shake just like a wet dog.

Horses are not random beings; there is always a reason for their actions.
So why do they roll?

Note: There are varying opinions on why horses roll- to stretch and loosen up, or maybe give themselves a back adjustment; maybe the mud helps keep them cool and deters biting insects from attacking them. But if you have ever watched a horse roll it's obvious that rolling is fun and feels good.

Stretch and Feel Good

Set the scene:
You've had a long and boring day at work. Most of your time was spent sitting at a computer.

You need a way to stretch your body and lift your spirits.

As humans we have devised many ways to achieve this goal and relieve the stress of of being cooped up all day.

For some people it involves saddling up their horse and going for a ride.
For others it may be yoga.

Human Equivalent

Interesting thought! We call these activities hobbies. I wonder if horses think of rolling as a hobby?

Set the scene:
The bay is a stallion and the chestnut is a mare.
The stallion gently reaches out to the mare with his head low.
He is very politely asking her permission to move closer.
In response she lowers her head and neck below his in a submissive gesture. Notice her ears are forward.

She is accepting his advances.

Note: If the mare had rejected his advances she would have jerked her head away, holding it higher with ears pinned back. She would also swing her body away from him giving him the cold shoulder.
This particular stallion was a gentleman and he would have understood exactly what she meant and pressed her any further.

Submission

Human Equivalent

Set the scene:
It's a beautiful afternoon. You decide to take a bag of stale bread to the local park and feed the ducks.

You bend down and stretch out a hand.

Even though we are higher up in the pecking order (excuse the pun) by lowering our body and reducing our height we assume a friendly submissive position.

And the ducks seem to understand.

Note - Adults will adopt a similar stance when talking to small children to reduce their height and be less intimidating.

Set the scene:
A large group of mares and foals are living together in the pasture.

Foals needs to learn how to behave respectfully in the herd.

Mares are generally very tolerant of all newborn foals but as they grow they are gradually taught herd manners.
Mares will first use "The Look" with laid back ears accompanied by a wave of the head in the foal's direction.
If a foal ignores the warning, the mare will follow up with a quick nip to enforce the message.
Mares will also use their rear end to warn a foal by turning their rump in the direction of the foal and lifting a hind leg.
Cheeky foals who ignore the subtle messages get a swift but non-serious kick.

Note:
All the mares in the herd (not just their own mother) are active in teaching the foals manners.

The Look

Human Equivalent

Set the scene:
Human mothers "All" know how to deliver "The Look".

The message is truly understood by the time children start school.

Kindergarten teachers can control a room full of energetic toddlers with just a wag of the finger and "The Look".

Do horses have an internal body clock?

Human Equivalent:

Think for a moment about your daily routine.
You probably wake up and go to bed around the same time every day. Do you eat your meals at roughly the same time each day? What about your hobbies and interests, are there set times when you ride your horse (perhaps after work).
Odd's are your personal hygiene routine also repeats itself at the same time every day.

Even humans who claim to be "free spirits" love routine, familiar patterns help us feel calm.
Putting order into our lives gives us the feeling that we are in control.
We force out need for routine and familiarity onto our domestic animals.
If you own a horse, chances are you feed and exercise him at the same time every day.
You may even give him the same day off every week.

These are very human traits, however horses quickly learn our routines and anticipate when it's either time to come in or dinner time.

Waiting

Set the scene:
The horse is standing by the gate waiting to be brought into the stable for his supper. He stretches his head and neck forward, opens his mouth wide closes his eyes and yawns.

Not once, but a series of 3 really big yawns.

Is he tired or bored?

Note: Some equine scientists believe when a horse yawns, especially when he emits several yawns in a row, he is relieving stress.

You often see a horse yawning after excersise, especially if the horse was learning something new or attempting difficult movements.

Tired or Bored

Human Equivalent

Set the scene:
You are in the library studying for an exam. Your goal is to spend 2 hours cramming. After about half an hour you start to feel stiff so you stretch and yawn.

Are you tired or bored?

Or stressed?

Whatever the reason yawning seems to help.

Note: In most cultures it is considered bad manners to yawn, especially in public, so we cover our mouths. If you are talking to someone and they start yawning, you feel they are not listening to what you have to say or they are bored with the subject and they are being rude.

Set the scene:
In the wild stallions are territorial. They mark their territory with what we call "stud piles".

The stallion will continally poop in one spot to create a pile, intensifying his scent and creating a marker that other males will recognize.

Despite hundreds of years of domestication and a firm understanding of how fences work, stallions still feel the need to create stud piles to keep other males out of their territory.

Note: Only the dominant stallion in the herd poops on the pile. There may be other males in the group but it would be considered extremely rude if they added their scent to the stallion's stud pile.

The offending male would do this as an act of defiance, challenging the stallion's authority and territory.

Territory

Human Equivalent

Set the scene:
You are on holiday.
Before breakfast you race down to the deck-chairs to place your towel on one to save your spot and ensure no one takes your chair.

After a leisurely breakfast you go back to the chairs to find someone has dumped your towel on the floor and is sitting in "your spot".

You are seething inside that they would be so rude when it was obvious that chair was taken.

However you have two choices. 1- Pick up your towel and look for another chair.
2- Confront the person sitting in "your spot" which may or may not go well.

Note: Humans are extremely territorial.

Bad Ass

A mean or belligerent person or
a person with unpleasantly extreme attitudes, behavior or appearance.

Note: Throughout my extensive research and a life-time of working with horses, the conclusion is that horses do not deliberately seek to harm humans. By nature horses are NOT mean or belligerent. Even when they are cruelly treated and abused they very rarely retaliate in an aggressive manner.
However there are exceptions to every rule.

Set the scene:

It's summer time, a group of brood mares are peacefully grazing in the pasture. Suddenly (unprovoked) one mare (the bay) launches an attack on another. The altercation is over quickly but the "bitchy" mare still appears upset and continues to flatten her ears and scowl at the others. There seems no reasonable explanation for the attack. There is plenty of food, ample room to find their own space and trees for shade and relief from bugs.

Her behavior may be hormonal, making it also irrational and beyond her control. She may be in season and ovulating causing fluctuations in her hormone levels. Perhaps she has already been bred and is pregnant. Just like humans some mares are more susceptible to PMS mood swings than others.

Bitchy

Human Equivalent

Set the scene:

"Bitchy," mean girls very rarely lash out with their fists but their words are capable of inflicting deep wounds. It would be nice if their bad behavior could be excused as hormonal or PMS, however humans are more complicated than horses. The reason for bitchy behavior could range from wanting to fit in with the "popular" crowd to self-preservation; hurt others before they hurt you.

Bromance is the combination of two words, "brother" and "romance".
A bromance is a close, emotionally intense, non-sexual bond between two males.

Geldings enjoy those close male relationships, for exactly the same reasons as humans.
* A “bromance” provides a like-minded, non-judgemental buddy to hang out with.
* Male (gelding) bonding can help them find their place in a mixed herd, feel more accepted and increase their feeling of belonging.
* A trusted friend who will have your back no matter what and someone to go down to the watering hole with.
* An exercise partner to gallop around and play with.
* It’s important to have a special friend to help you get to those hard-to-reach itchy spots (mutual grooming).

Mares may also form very strong friendships however a “bromance” is not effected by hormonal fluctuations.

Chillaxin

A mixture of Chilling out and Realxing.
Something horses are extremely good at.

Dawdle

To move slowly and waste time

Set the scene:
You are calling your horse in from the pasture. Usually he is happy to see you and gallops across the field as soon as you call his name.
Today he would rather hang out in the pasture with his buddies; he walks as slowly as he can to the gate. Still obedient but zero enthusiasm.

Human Equivalent

Set the scene:
You are nine years old and have enjoyed a glorious summer, getting up late, playing with your buddies all day. No work just play.
But it's September and time to go "back to school".
You walk to school as slowly as possible,
dragging your backpack behind you,
trying to savor every last ounce of freedom.
You know you have to go,
but you really don't want to.

Goofing Around

Playful (non-aggresive) displays

No one knows for sure if horses have a sense of humor, if their occasionally silly antics are intended to amuse and entertain their buddies. However if you spend long enough just hanging out in the pasture with a herd of horses you may come to the conclusion that perhaps they do. Otherwise what would be the point of their goofing around?

Set the scene:
Ponies are notoriously greedy and the only thought on this pony's mind is eating grass.

His efforts to put his head down and graze are thwarted by the "grass rein" device attached to his bridle.

Grass reins helped the young rider complete her dressage test. However when they halted at X to salute the judge, the pony frantically pawed at the ground desperately trying to bite at the grass that was so tantalizing, but out of reach.

Hangry

When a lack of food causes you to become angry and frustrated.

Human Equivalent

Set the scene:
Only fueled by coffee for breakfast, you tackle your morning chores. Then you get a phone call from your child's school, your child is sick and you need to come and pick him up. You drop everything and race to the school. Before you can head to the doctor's office with your sick child there are release forms to sign but the secretary is on her break so you have to wait. Finally you head out to the doctor's office, only to sit in the waiting room for over an hour. Finally you see the doctor, pick up the prescription and drive home. Your tummy is grumbling and you have a hunger headache. Suddenly a car comes out of nowhere and cuts you off. That is the final straw, you loose it, you thump on your horn and shout abuse at the other driver (even though he can't hear you). Your shouting upsets your child and he starts to cry which makes you even more irritated.

You are not yourself when you are hungry.

Set the scene:

Whether you are a fan of the rodeo or not, you have to admit the bucking bronc has the shortest work day of all riding horses. His shift only lasts 8 seconds before the bell rings and it's quitting time.

The bronc is not happy to be saddled and even more upset that someone tries to sit on his back. This is the classic example of a horse throwing a "hissy fit" because he is not getting his own way.
Even after dumping his rider, he is still mad.

Note: Bucking broncs are not rank, nasty horses. Once the saddle is off and he is back in the pen, the horse instantly calms down and hangs out with his buddies (the other bucking broncs).

Hissy Fit

A sudden outburst of temper, originates from an abbreviation of "hysterical fit".

Human Equivalent

Although you immediately think of small children throwing temper tantrums because they don't get their own way, emotional explosions have become common place in professional sports. From tennis to football athletes throw a hissy fit when the umpire or referee makes a decision they don't agree with.

Humans are more prone to hysterical outbursts. In fact you would only have to follow a parking enforcement officer for a day to get your fill of adults throwing "hissy fits". From mild as they rip the parking ticket from the windscreen, to full blown as they see their car being loaded onto the tow truck to be impounded.

Set the scene:
Two new mares are introduced to an established large herd.
After the initial introductions of sniffing and squealing the two new-comers wander off together and are grazing peacefully, until the grey gelding decides to impress them with his good looks and fancy moves. He puts on quite a show for the girls, which engages their attention for a while. However they quickly realize he is a gelding and all the fancy moves in the world won't make him a stallion. Unfortunately the gelding has no value to mares; he is neither the Alpha of the herd or a possible mate.

Jocks

Urban slang for good looking popular boys, usually on high school sports teams.
They have an inflated opinion of themselves and are convinced their athletic abilities make then better than everyone else. Their popularity and status rarely carries on after high school.

Loud

Attention seeking

More of a behavioral trait than body language, this may manifest itself in a variety of ways, from calling out to other horses to banging on the stable door.
The horse needs to be seen and heard by others, probably due to insecurity issues.
They seek validation and attention even if it is negative attention.

Human Equivalent

Someone who likes to announce everything in their life on Facebook to get people to feel sorry for them or get a reaction. Their stories are either grossly exaggerated or completely fictitious in order for people to respond to them. They crave likes and comments, it doesn't matter whether they are good or bad. They just need to know someone is listening.

Set the scene:

Each horse was given his own bucket containing his lunch. The buckets were spread apart to prevent any squabbling.
The white horse in the middle is old, and a slow eater. When the younger horses had finished their food they all came to "mooch" off the old grey mare. Although she may not have been thrilled about having to share her lunch, she allowed the others to eat her food.

As soon as the food was all gone the young horses wandered off leaving the grey horse alone.

The picture above is a very unusual scenario as horses don't usually like to share their food, in fact some get extremely aggressive if another horse attempts to take what is theirs.
This mooching behavior is only taking place because the grey horse is allowing it.

Mooch

Human Equivalent

Set the scene:

We all know someone who fits the "mooch" description: the cheque comes, and they realize they have no money. They come over for dinner, usually late and empty handed, but leave with leftovers and a bottle of wine. You always drive but they never offer you gas money.
You may be related to this person or they might be a friend. You love them but they drain your resources and patience.

Although we are talking about the human equivalent here, I bet you know of horses who act in a very similar way.
They nuzzle your pockets for treats and when the treats are all gone they run back to their herd buddies. They invade your personal space and push you out of the way to get where they want to be.

"Mooching" only works if we let the mooch get away with such rude behavior.
The key is to set boundaries.
Before you can set boundaries, you will need to figure out what you are comfortable with and what you will not tolerate.

The only way a horse can change his appearance is by rolling in the mud. I am not sure if their intention is to look "stylin" or they deliberately want to change their natural coat color. However it seems a very popular pastime especially with horses who have just been groomed or bathed (much to the dismay of their owners).

Horses roll in the mud for a variety of reasons. For example, rolling is a great tool for horses when they want to self groom, you will notice horses who have just been ridden, like to roll. Maybe to relieve the itchiness of a sweat-mark made by the saddle.
Rolling; is how a horse scratches his own back.
Horses like to roll around after a bath because as the horse's skin and coat dries it becomes tight and itchy.
The action of rolling may help to ditch the itch.
Rolling in mud is an added bonus, the mud works as a sunscreen as well as providing a barrier between the horse and biting bugs and insects.

Stylin

Looking good or in fashion

Men and women are constantly looking for ways to change their appearance. However their motivation it to gain the admiration of others and has no practical application.

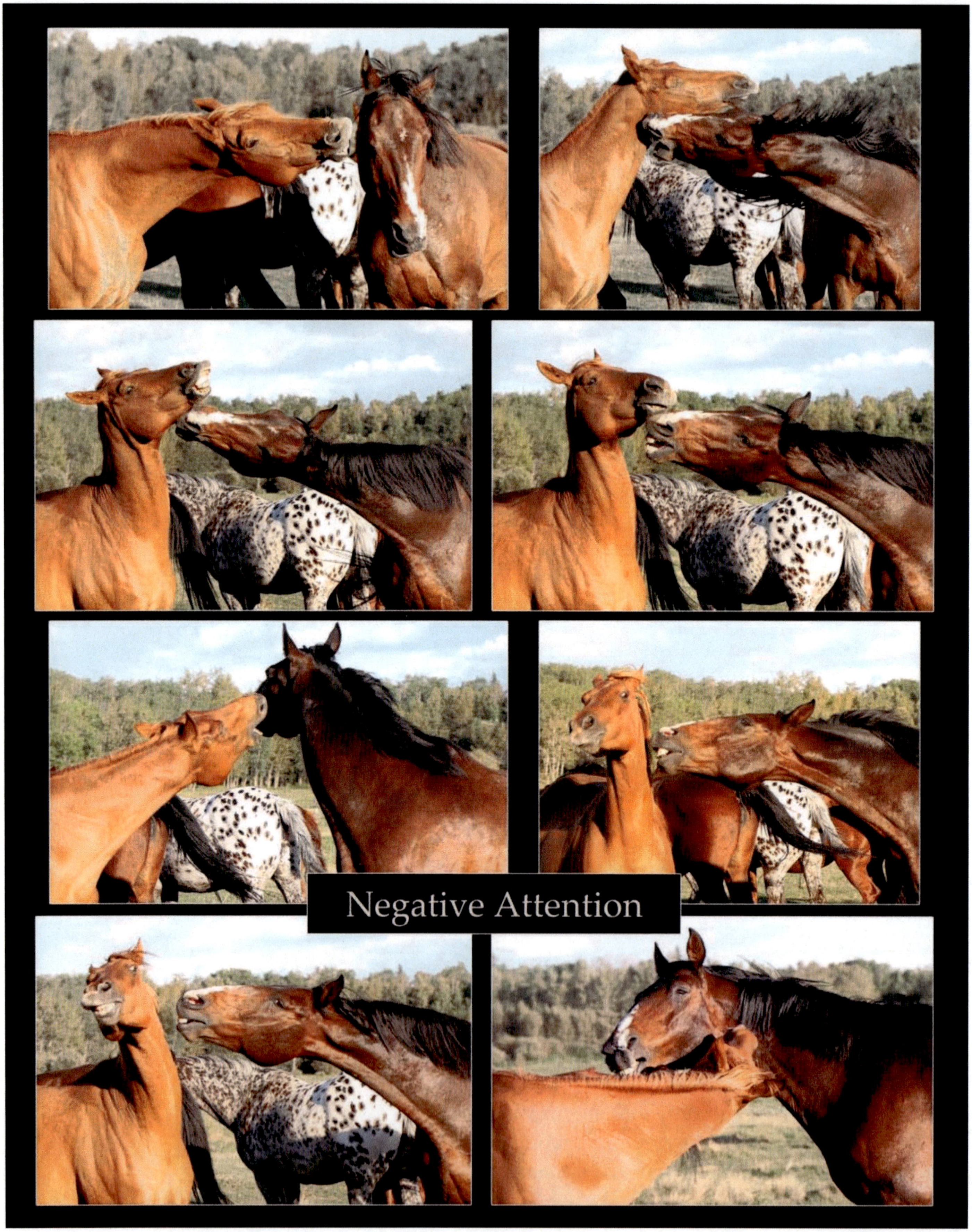

Negative Attention

Negative Attention is Better than no Attention

Set the scene:

It is summer time, the herd is grazing peacefully, all except one horse who appears bored and is harassing another horse (for no apparent reason). The bay nips at the chestnut mare, not in a particularly aggressive way - just to get her attention. She nips him back as though to say "knock it off". However the bay got the attention he was seeking so continues to nip at and annoy the other horse. This went on for a few minutes until the chestnut mare whipped her back end around to give him a swift kick before running off. Undeterred the bay galloped after her.

Needy

Need attention whatever the cost

Human Equivalent

Think back to your school days. There was always that one kid who would disrupt the class. "The class clown", who would crack jokes in class, talk back to the teacher and goof around. He distracted the other kids as they laughed and giggled at his antics. The teacher had her hands full trying to control the class clown. He had a special seat at the front of the class; he was a regular visitor to the principal's office and very often stayed late in detention.
Even the punishment provided him the attention he desperately needed.

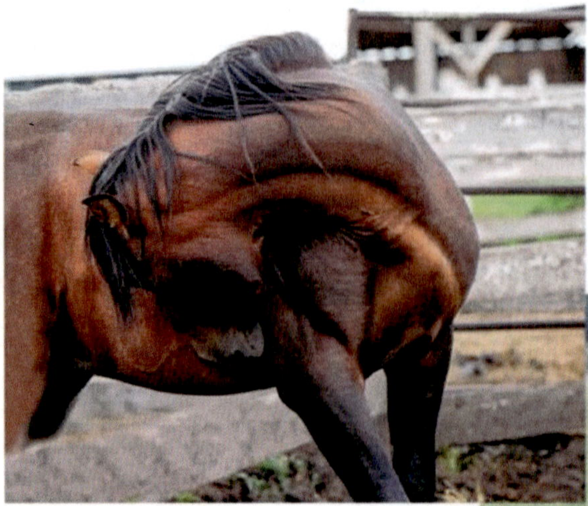

Set the scene:
Sometimes a horse will get an itch they just can't seem to reach. However they are smart enough to figure out ways to ditch the itch using a variety of inanimate objects as scratching tools from trees to fence posts.

Smarticle

A smart / intelligent person

Human Equivalent
There is a new movement that is both fun and informative.
Created by some smart and innovative people who have devised ways to use everyday objects to solve annoying problems.
If you search "Life Hacks" on the internet you will be amazed at the fascinating ways they have found to make life better.

Skedaddle

To leave a place quickly

Set the scene:

You head out into the pasture to catch your horse. He is usually easy to catch and occasionally he will come to you when you call his name. However today when he hears your voice and sees you approaching, he runs off. This has nothing to do with "Fright and Flight". The horse is not scared. He has made a conscious decision to stay in the pasture and have a play day.

He knows he can out-run you so all he has to do is skedaddle when you approach.

His body language is sending a very clear message.

The reason behind his action may need further investigation.

Human Equivalent

A couple of kids sneak into an orchard to steal fruit from the trees.

The farmer spots the intruders and yells at them get out of there before he calls their mother.

The kids pick up their stolen apples and skedaddle.

Instinct versus Cognitive Awareness
Game changing insights

Instinct: An instinct is something you don't need to learn, it happens naturally, without you even thinking about it, like remembering to breath in and out.
It's instinct that tells the birds when to begin their migration.

Cognitive Behavior: Refers to things one does to achieve desired outcomes. Like the remember-and-response sequence (do things in order) etc. Recognizing the consequences of their behavior or actions is thought to develop as people mature. With age and experience they can envision what might happen. Cognitive behavior is thinking behavior.

A popular belief among scientists is that humans beings are the only species who are evolved enough to make cognitive decisions and that most of the animal population works on instinct. They do waver a little when faced with exceptional animals who have been taught to do exceptional things. However there is always a somewhat arrogant feel to their findings as they conclude that certain individual (not all) animals may have greater abilities than the rest of their species, but only because the talents and tricks they perform were taught to them by a human.

Some (not all) horse trainers share the view, "The horse is only as smart as the person who trained it".

If you believe that horses act purely on instinct and are not capable of independent (not taught by humans) cognitive thought and decision making, this chapter will blow your mind. The studies, stories and visual proof that horses are not only capable of independent decision making, but they also have the ability to predict future outcomes based on their actions (abstract thinking) will shatter the illusion that horses act purely on instinct.

If this concept alone was not enough to blow your mind, over the course of the next few pages you will see proof that horses are also capable of a much deeper trait. It is a quality humans struggle with, a social skill that has to be repeatedly taught to toddlers so they will be accepted into society.

"Sharing is Caring"

The life lessons we learned in kindergarten to share the toys and to "take turns" even though we didn't want to, were an essential part of our education. They were not popular lessons and some kids took longer to conform to this social nicety than others. As adults we know we need to hide this selfish trait even though society admires the ones who "have it all" or climb over everyone else to get to the top.
Likewise there is a fascination for reality TV shows that depict spoiled selfish people having tantrums to get their own way. Is it that we secretly admire their disregard for social norms as they act like a toddlers and put their own needs above everyone else? Or are we eagerly anticipating the negative outcome, when they get their just deserts from their bad behavior?

In Chapter Two we talked about Maslow's hierarchy of needs and discovered exactly what motivates horses to do the things they do.

Equine Hierarchy of Needs

Food is the number one motivator for most living things including horses.

Set the scene:

It's quite common to see horses peacefully eating together at a round bail. After all there is plenty of food, no one will go hungry so this is not really sharing, just an example of calm cooperation.

But what if the hay is limited and there is not enough to fill the bellies of everyone?
This is an easy experiment - one slice of hay was put out for 4 horses. It was deliberately left in a small pile and not spread out to allow all horses easy access.
Some may think this a cruel experiment - but the horses were not fazed by the apparent shortage of hay, in fact they happily munched on the hay together until it was all gone.

Hmm. Not really the result one would expect so let's up the ante and put out one small bucket of oats for 4 horses. The stakes are higher and the prize is greater.
However the results are the same. The horses considered the needs of the others to be equal to their own and shared the limited grain. Not one of the horses tried to steal all the grain for himself.

On first blush you may not think this a such a big deal, and to be honest neither did I. After all, these horses knew each other, they obviously got along with each other and probably ironed out any kinks or disputes about sharing a long time ago.

Until the thought that perhaps horses are capable of "unselfish acts" occurred to me.

Unselfish: *Willing to put the needs or wishes of others before one's own.*

That was a mind-blowing thought, but even more incredible, was that it's true. Horses are capable of and in fact exhibit many examples of unselfish behavior, incidents when they put the needs of others before their own. They do it with other horses and they even extend this act of altruistic behavior to other species including humans. While conducting environmental observations and research for this book I had one basic rule, "If I can't photograph the behavior or action it must be just hear-say or myth". After all, seeing is believing. This meant conducting many, many equine experiments to test my theories and challenge popular (but un-proven) beliefs.

The experiment was conducted again this time with just two horses that didn't know each other and what took place over the small bowl of food was truly beautiful to watch. A small bowl of grain was placed between the two horses.

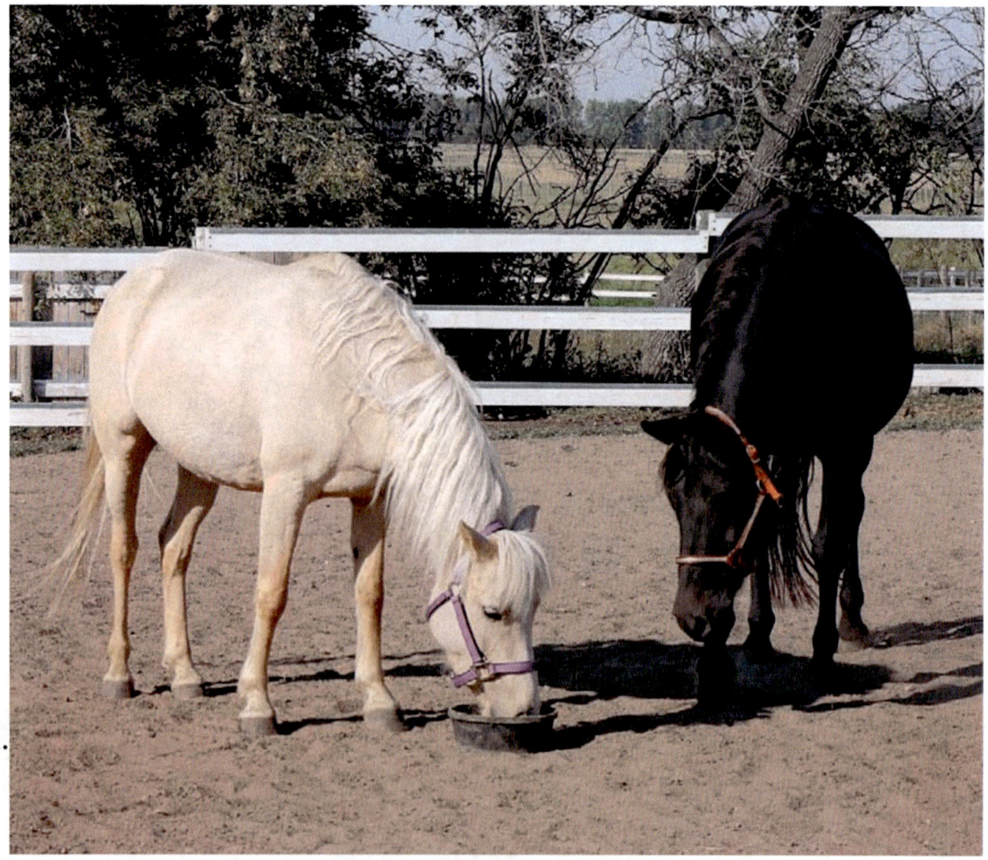

The palamino was first to put his nose in the bowl and started to eat the grain.

The black horse slowly approached the bowl. He obviously wanted to join in.

The palamino lifted his nose from the bowl and turned his head and neck to the side to allow the other horse to eat.

After a few moments the palamino returned his nose to the bowl.

The black horse mirrored the sharing behavior by lifting his nose from the bowl and turning his head and neck to the side.
As if to say, "Go-ahead it's your turn now".

They repeated this "sharing behavior" until all the grain was gone.

Let's move to the next rung on the hierarchy ladder. Once the horses have food, air and water they seek **Safety and Protection from Predators**.
Here too there is evidence of unselfish behavior.

Set the scene:
Consider the basic herd (or family unit), a stallion and his band of mares and youngsters. The stallion is easy to distinguish from the rest of the group as he always appears to be busy. The main group seems relaxed as they graze in the pasture but the stallion continually circles his family keeping a keen eye on the horizon looking for danger in the form of predators. He repeatedly sniffs the poop surrounding his family so he knows exactly who is present (he can identify each horse by the unique smell of its poop). He is checking for marauding bachelor studs who might try to sneak up and steal his mares.
Some may say this behavior is instinct and he is hard wired to act this way.
Maybe there is some truth in that, however horses are also hard-wired to flee in the presence of danger. This means by virtue of his continued vigilance the stallion would be the first horse in the group to see a wolf or mountain lion.
If he was acting on instinct he would panic and bolt to safety.

In reality he suppresses his own natural fear of the predator and moves the herd to safety. He puts the safety of his family above his own.

A stallion moves his herd by lowering his head and neck and moving it from side to side in a very distinctive snaking movement. Take note of how all the mares comply to his wishes - when he says move, they all move. They know he has their best interest at heart.

There are stories of wild horses turning around and attacking wolves in order to buy some time, allowing his herd to escape. This is a very romantic story and would be lovely if it were true. However I have never seen or photographed a stallion attacking a predator, and in my book "seeing is believing".

There is plenty of documentation and photographic evidence that a stallion will attack and chase off young studs who threaten to either sneak into the herd and breed with one of his mares, or worse try and steal a mare for himself so he can start his own family.

This sequence of images shows "wild" bachelor stallions fighting over a filly.
They managed to steal the filly from another herd. The young studs will squabble and fight over her until one stallion wins the spoils and leaves the bachelor group with his filly to start a band of his own.

Exception to the Rule

Hopefully you are coming to realize the rules concerning instinct and herd behavior for the most part are sound and true. More investigation reveals interesting findings that cannot be ignored. For decades researchers focused on finding general patterns to describe and label natural instinct. Once they found a general connection they moved on to the next. There are always exceptions to every rule, but what if the "exception" is a more accurate representation of natural behavior and the rule needs to be re-examined.

Before we move on I want to share a little story about a very unusual breeding farm.

Polar Pintabians is a breeding facility owned by Mr. Bob Williams whose objective was to produce top quality show/performance horses. For many years he bred beautiful horses and enjoyed considerable success.

Mr. Williams followed the same general procedures most breeding facilities set in place when managing large numbers of stallions, brood mares (visiting mares) and young stock.

His stallions were kept in their own large pens away from the rest of the herd and fed grain and hay. Brood mares and their young were allowed to graze in large pastures until they came into season, at which time they were brought into the yard to be "hand bred".

(A groom would lead the mare in front of the stallion to arouse him. When he was all excited and ready to breed, Bob would bring the stallion out of his pen and hold him while he covered the mare.)

Hand breeding is a very efficient method as it gives horse owners complete control over which horses mate, ensuring the horses create blood lines that in the horse owner's opinion will produce the best foals.

Bob owned and cared for a large herd consisting of eleven stallions and over fifty five mares and youngsters. Although the breeding model had worked extremely well for many years something bothered Bob about the "set up".

"I had a couple stallions that had never been out of their pens and in a pasture which meant they were eating out of a bale all the time, and were developing breathing problems. I wondered how they would react if I turned everyone out into the 100 acre pasture all together. Initially I worried that the stallions may fight and hurt each other. Strangely enough it never really happened; there was some initial bickering but nothing serious. They had a few bite marks, but after a few days all the disagreements blew over."

The next chapter in this little story is truly fascinating:

After only a few days Bob noticed the herd splitting up into distinct and separate groups. They had formed mini family units consisting of one stallion and a few brood mares and her offspring. Six of the eleven stallions had managed to create for themselves their own little harem. It is worth noting that the groups were not evenly split, some stallions had more mares than others. The smallest group consisted of a stallion and just two young mares. However once the groups were established no squabbling or fighting took place and they stayed together. The horses enjoyed the freedom of 100 acres of grassland. If one band moved too close to another band the stallion would just quietly move his mares away. They acknowledged that there was plenty of room for everyone to live in harmony.

Bob's experiment played out extremely well - his stallions were enjoying a natural lifestyle, fresh air, plenty of exercise and nature's food and very quickly their respiratory problems disappeared.

The chestnut stallion secured two black mares for himself at the beginning of the experiment and sired two foals.

The black and white stallion did considerably better at gathering a group of mares. It's interesting to note he was attracted to and included in his harem mostly paint mares.

This black and white stud also selected only paint mares to add to his family and breed with.

The buckskin is the stallion in this family group.

Notice how the lead mare lowers her head and neck in a friendly, non-threatening gesture as she approaches another family group. The stallion (dark bay horse at the back of the group) gently pushes his family forward.

You may be wondering about the other stallions who didn't manage to acquire a mare or two and create their own family unit.
They banded together in a bachelor group and hang out together in their own little tight-knit band. For now they seem to have resigned themselves to a celibate lifestyle.

Although Bob's horses were all born and bred in captivity and enjoy human contact they are enjoying the same lifestyle as wild horses. They have freedom to form relationships with the horses of their choosing. It was very interesting to observe how the groups were formed. These horses appeared to have a preference when it came to choosing their herd mates; they favored other horses that looked like themselves.

Human equivalent
Have you ever noticed that when you travel abroad on holiday and by chance find yourself talking to someone from your home town or country you instantly like them? You enjoy chatting with them to find out if you have similar friends or are connected in any way. We feel comfortable when we have found our own tribe.

I hope you enjoyed the little side story of Bob's Pintabian experiment, but lets turn back to the third rung on the hierarchy ladder which is the need for family and establishing your place in the herd.

In this area there are several indications that horses not only acknowledge the needs of others they also put them ahead of their own.

The most beautiful example of this is when mares baby-sit foals while they are resting. In the picture you will see several mares all standing around (protecting) the foals who are snoozing. Even though there is no immediate danger to these domestic horses, the mares put aside their own need to move around and graze to watch over the foals and keep them safe.
Note: There are more mares than foals which means not all have a maternal connection to the foals.

It's not just brood mares who exhibit this act of service.
These yearlings take turns. A few will keep watch while the others take a nap.

We could, of course, talk about the many ways mares care for and protect their young, although these are examples of equine behavior more than body language. They may also be bordering on instinct rather than a cognitive unselfish act of service.

In this chapter we are searching for evidence of this cognitive social skill.

Which brings us to the fourth rung on the hierarchy of need pyramid: Esteem - acquiring skills that lead to love and respect from others. When all the basic needs of domestic horses have been met they seek love and approval from their human owner.

Esteem

Acquiring skills that lead to love and respect from others

This has to be the most unselfish act of all. The horse strives to understand the demands made of him in order to please his owner. These demands may mean jumping over huge fences, spinning and performing sliding stops or achieving collection and extension of his natural paces. These skills add nothing to the horse's self-worth. He has no personal need to excel in these disciplines or win ribbons. His motivation and reward is the love and appreciation of his owner.

One more example before we move on.

You may have heard stories about how usually high-spirited and nervous horses are quiet and docile when there are young children or people with disabilities around them. It's as if the horse knows he must be extra careful not to harm them in any way. This happens more often than you may imagine. However I want you to ponder this scenario for a moment.

The horse is able to modify his true nature to accommodate the needs of someone who obviously cannot cope with his authentic highly-strung personality. He consciously calms down to care for the well being of another - and they are not even the same species.

An off-the-track Thoroughbred quietly babysits this young rider. Yet when a more experienced older rider climbs on his back he is high-spirited and eager to go fast. This is clearly NOT instinct but a deliberate, conscious effort on the part of the horse to put the needs of his young rider above his own need to run.

Why Horses will "Do Tricks"?

When you think of how our body language relates to and is interpreted by horses you may initially think of horsemen and women who teach their horses to do tricks.
They use mirroring techniques to not only train but, cue their horses what to do.

This book is not about trick training, in fact it has very little to do with horse training of any kind. The purpose of de-coding equine body language is to deepen your relationship and have a greater empathy for horses in your care and those you come in contact with.

Trick trainers are extremely in-tune with how they nonverbally communicate with their horses. For a few lucky individuals this seems to be a natural ability, which I felt was the case with a very special little girl called Carlie Borle.

Carlie, at just 11 years of age, has personally trained four of her Shetland ponies to perform a variety of tricks, including "Roman Riding". She stands atop the pair with one foot on each pony and rides them through barrel patterns, performs figure eights at walk, trot and canter. She even jumps them over small obstacles. Carlie takes her spectacular show on the road to equine events and rodeos where she entertains the crowds. She is not only super talented but also super cute.

This little girl and her ponies intrigued me so I set up an interview with Carlie to find out how, at such a tender age, she managed to accomplish such amazing equestrian feats.

You may be surprised at the answers.
Q. How old were you when you started riding and training ponies?

Carlie - I started riding when I was five years old and I competed at horse shows in the hunter and jumper classes.

Q. What inspired you to teach your pony tricks?

Carlie - I enjoyed teaching my dog to do tricks so I thought maybe I could teach my ponies the same way.

Q. Who taught you how to train your ponies to do tricks?

Carlie - I taught myself and watched a few YouTube videos for tips and ideas.

Q. What was the first trick you taught your pony and how did you do that?

Carlie - "The Bow" was the first trick I taught my pony - I just kept pulling one front leg out in front of him and encouraging the pony to put its head down using treats. My ponies are really greedy and will do most things for treats and affection.

Q. What was the hardest trick to teach and how did you overcome the problems?

Carlie - "The Rear" was one of the hardest tricks. To get the pony to lift its front legs off the ground I stood on a box and used treats to make him jump up - just like I did with my dog. When he jumped up he got the treat and lots of praise, it took about 3 weeks to get him to go all the way up into a full rear. Now all I have to do is lift my arm and he knows what I want him to do.

Q. If your pony really doesn't want to do a trick how does he let you know?

Carlie - If he doesn't want to play or he has had enough he gives me "The Look", flattens his ears and looks really grumpy. So then I do something else with him - either go for a ride or groom him.

Q. Do you think the ponies enjoy performing?

Carlie - Yes, for sure. I have put our routine to music and when they hear the music starting they get excited and want to begin. They love the crowds and love to show off.

There are lots of books about trick training - in fact there are lots of books about "how to train" horses in a variety of disciplines. The reason I included Carlie's story is because she was an inspiration to me. I had been struggling with how to explain and show how horses mirror both humans and each other. However whenever I thought of mirroring my thoughts would always come back to trick training. It was like I had hit a brick wall.

So I took a lesson from Carlie and instead of looking for complicated adult solutions and observations, I approached the subject of mirroring with a child-like simplicity.
By asking the question, do horses "Mirror each other and if they do - why?"

Mirroring Behavior

Why all nammals engage in mirroring

What is mirroring and why should horsemen/ women concern themselves with it?

According to physiologists:
Mirroring is a subconscious behavior in which one person imitates the gesture, body language, speech pattern, or mood of another. Mirroring takes place in everyday interactions, both at work and socially. Generally it goes unnoticed by both people. If harmonious mirroring took place both people will feel like the interaction went well, that they really had a connection and mutual understanding. They may believe that they share similar attitudes and ideas as well.

Humans learn this nonverbal method of communication at a very early age. We all know how easily kids will pick up (copy) our bad behavior because they want to be just like us. Unfortunately we try to educate our kids to "do as I say" not "do as I do".
If you exhibit road rage and yell at a driver who cuts you off, don't be surprised if your toddler, sitting in his car seat in the back, also gives the driver the finger. Perhaps you drink and smoke yet tell your kids cigarettes and alcohol are bad for you. Don't be surprised when you catch them behind the garage smoking and drinking booze from your liquor cabinet.

As we get older we use mirroring body language to show empathy.

For example, if your friend came to you weeping and heartbroken because her boyfriend dumped her, you unconsciously adopt the same downcast demeanor as you try to console her.

Perhaps another friend tells you she just got a promotion at work and wants to celebrate. You quickly move into an upbeat mode, smiling and mirroring her excitement. Unconsciously you want to show you are happy for her.

Romantic couples provide a fabulous example of mirroring. If you ever go to a populated public area, like a park or outdoor cafe, you'll often notice that couples are mirroring each other as they interact. Very often they will walk in step with each other, lean in to each other when talking and copy each other's gestures.

Mirroring even takes place in large group settings. Think how quickly the "wave" happens at a football game when thousands of individuals become a single unit. Or how large groups of peaceful demonstrators can quickly turn into an angry mob.

As humans we adopt this nonverbal form of communication for three reasons:

1. To show empathy
2. To let the other person know we understand (on the same page)
3. To form a connection to the other person and build rapport

So what has this got to do with horses?

Did you know horses are experts at reading our body language? In fact they can read your body language from across a field.
In recent years a new breed of horse trainer has emerged. Talented horsemen and women who have learnt how use body language and mirroring techniques in the popular "Natural Horsemanship" method of connecting with and training horses. Using body language and gestures they can train horses to follow (mirror) their movements. Working with horses from the ground (so the horse can see them) they are able to execute a wide variety of movements and tricks.

The main reason the natural horsemanship technique is so successful is because horses "mirror" humans for the same reasons that humans mirror each other:

1. To show empathy
2. To let the other person know we understand (on the same page)
3. To form a connection to the other person and build rapport

However for every Ying there is a Yang.
For the record, horses are also able to mirror bad behavior and negative emotions just as easily.

Set the scene:
You had a bad day at work, got stuck in traffic on the way to the barn, only to find when you went into your tack room there had been a waterpipe burst and all your gear was sitting in puddles of dirty water. Understandably you are feeling angry and frustrated. Just to add to your frustration, your horse that is usually easy to catch won't come near you and forces you to chase him all over the field. Eventually you catch him and tie him up to groom him before your ride. You wonder what is wrong with your horse - he is antsy and fidgety. You respond by yelling at him to stand still but he doesn't listen. When you eventually climb into the saddle to go for a ride, he won't go forward; there is nothing for him to be scared of but he spins around in circles not quite knowing where he wants to go.

As you read through this scenario it's easy to see how the mood and body language of the rider effected the horse, which he in turn mirrored back.

This is where the myth that horses can smell fear came from. There is no scientific proof that fear has a certain scent, however horses can easily spot the body language of someone who is nervous or afraid. And guess what, the horse wants to please and form a connection so he mirrors that behavior.

Note: You can mask your fear by singing or humming a tune - this helps to regulate your breathing and helps your brain to think of something other than what is making you nervous.

The purpose of this book is not to promote or teach natural horsemanship nor is it concerned with how to train horses. The goal is to better understand equine body language.

Did you know horses mirror each other too? As far as we can tell for similar reasons as humans:
1. To show empathy
2. To let the other horse know he understands how he is feeling
3. To form a connection to the other horse and build trust

One horse hears something in the distance and picks his head up and looks towards where the sound came from. He pricks his ears in the same direction to catch any extra auditory clues as to what might floating in the wind.

The horse standing next to him mirrors his behavior and looks in the same direction.

Human equivalent
If you stand in front of a tall building and look up and point in the direction of the roof - after only a few moments other passers-by will stop and look up. Some will even point without even knowing what they are pointing at.
Is this just curiosity or is it mirroring?

The first horse decided there is nothing to look at after all and resumes grazing - likewise the other horse mirrors that behavior too.

You might be thinking this was just coincidence, and perhaps it was. However if you spend some time in a pasture "Horse Watching" you will see lots of examples of horses mirroring each others' behavior. It can't all be purely coincidence.

Stampede is a classic example of mirroring.

One out, all out!

What could possibly make the grass just under the barbed wire so appealing?
It certainly doesn't look anything special.
Maybe the first horse knew something the others didn't so they decided to mirror his behavior and join in.

Hmm. What are the odds of four horses needing a drink at exactly the same time?

Some horses will get down and roll at the exact same time.

Very often when one horse yawns others will follow suit (this happens with humans too - yawning may just be contagious).

I hate to bring this up but, back in the day when I taught horse riding, if one horse decided to poop, they all decided to join in. Maybe they thought it was funny to watch me frantically scooping up the poop.

Mutual grooming could also be considered a form of mirroring behavior. Horses often begin a mutual grooming session by scratching each other's withers but then move up and down each other's body using their teeth to scratch and gently nip their grooming partner. It is thought they show their partner where they would like to be scratched by biting the mirrored spot on the back of the other.

Perhaps this is where the saying "You scratch my back and I will scratch yours" came from.

Mirroring

Weaning Issues

Do horses comfort each other?

In times of trouble do horses comfort each other?

This might seem like an unusual question, but one I found intriguing.
Humans have a wide variety of ways they show compassion and concern for another's discomfort and pain, from a simple hug that shows you care to providing meals (kindness) to a sick friend.

Do horses exhibit behavior and body language intended to comfort another horse?
To answer this question, first we need to look at a situation where several horses are experiencing the same amount of discomfort and stress. Only then can we document their actions to determine if they comfort each other and if they do "what does that look like"?
Set the Scene:
We are at a large horse breeding facility on a beautiful sunny day in late August.
The mares and foals are grazing peacefully in the pasture unaware that today is "weaning day". Most of the foals are around 3 months old and are getting all their nourishment from grass - they will still suckle from time to time but at this age derive very little of their daily food intake from their mother. The youngsters are becoming more independent and will wander away from their mother to frolic and play with the other foals (their buddies), however quickly run back to their moms if they get spooked or sense danger.

It is customary for breeders to wean all the foals on the same day in the hopes that company will ease their stress and help them with the transition.

The weaning pen was a large high-sided pen, designed to safely house the foals for the next couple of weeks. There was a round bail in the center and fresh drinking water in the corner. A mare accompanied by her foal was led up to the pen, the foal was ushered in and the mare led back to the pasture.

Foal number one came into the pen.
He was obviously very distressed; he was whinnying (calling out) to his mother, continually running up and down the fence looking for a way to get out and back to mom. It was evident from his body language that he was extremely upset. As he ran his head and tail were held high and alert, his poop was loose and more frequent than normal, plus he showed no interest in eating or drinking.

Hopefully when foal number two is added to the pen he will draw comfort from his pasture buddy. After all they are both in the same boat.

Foal number two is introduced to the holding pen.
His actions mimic the first foal, calling out to his mom and frantically running around looking for a way to get out.

The foals barely acknowledge the presence of each other - until in their frenzy they accidentally bump into each other.

The result of the observation was truly disappointing. Instead of comforting each other, which is what I secretly hoped would happen, they scowled and kicked out at each other, as if annoyed that the other foal had touched them. Instead of drawing comfort, the presence of other foals seemed to add to their stress and irritation. They just wanted their mother, not other foals.

This pattern repeated itself as each new foal was added to the pen.

The mares showed little or no concern that their babies had been taken from them - they went back to grazing in the pasture as if nothing had happened.

To fully understand how the foals are feeling consider this human equivalent:

Set the scene:
A mother and her toddler go into a department store. The mother has a tight hold of the child's hand but in a moment of distraction the toddler slips loose and runs off.
The mother is frantic. She is calling her child's name and running up and down the aisles looking for her run-away child.

The child thinks this is a great game and is playing hide-and-seek amongst the racks of clothing quite oblivious to the stress she is causing her mother.

A sales assistant touches the mother on her arm and asks if she can help her. Mom is totally un-glued by this time and brushes her off screaming that she has lost her baby and has to find her. She continues calling out and running up and down the aisles. Eventually the toddler pops out from under the garment racks giggling.

Any mother reading this will be able to identify with that awful gut-wrenching feeling when you think you have lost your child.

Walk the Walk

Surprising insights from observing this simple gait

Walk the Walk - Talk the Talk

This is a very common phrase that I am sure everyone has heard, but have you ever considered the literal meaning of this little ditty?

Walking is something we put very little thought into - after all it's just placing one foot in front of the other. We focus on where we are going rather than the style or pace at which we are walking. But how fast you walk and the style in which you walk speaks volumes about both your personality and intent.

This is a short list of different walking styles. There are many more but for now we will just look at styles that we can compare to how horses walk.

Amble - A slow leisurely walk

Limp - To walk slowly with some form of physical injury

Lumber - Move heavily or clumsily

Lurch - Move forward as though you have little control

Meander - Walking slowly, not on a straight course, as though unsure of where you are going

March - This style of walk requires much thought and deliberation to keep in step - a common military drill practice

Parade - Walk in procession one behind the other all keeping the same pace

Prance - Walk with an animated light step

Prowl - Move in a predatory manner

Saunter - A slow aimless walk as though you have no real destination, one would saunter through a park, taking time to smell the roses

Shuffle - Walk slowly, taking small steps and pushing your feet along the floor rather than lifting them up, giving the impression that you are in NO hurry to reach your destination

Stagger - Walk with great difficulty, unbalanced and unregulated steps “staggering” as though drunk

Stalk - Walk stiffly as though ready to pounce

Stroll - Nice leisurely pace, with some direction but no urgency to get wherever you are going

Stomp - Walk heavily, trying to put as much weight as possible into each step

Stumble - Miss a step and nearly fall

Swagger or Strut - To walk with a confident sway in the hips

Tip Toe - To quietly sneak up on something or someone by taking small steps and walking on the balls of your feet

Trudge - Slowly drag one foot in front of the other with great effort

Waddle - OK this one is more a description of how a baby duck walks, or used as an insult

Wander - Slower pace with little or no real direction

Walk Briskly - A fast walk with a clear intention of where you are going, you are on a mission

Your style and pace of walking reveals more than you think.

To test this theory:
Go to a side walk cafe, sit for a while and “people watch”. You will see examples of pretty much all these walking styles within a very short period of time.

You will be amazed at how easy it is (when you take the time to watch and think about it) to read their body language by the way they walk.

The guy in a suit, striding along at a swift pace while talking on his phone sends a completely different message from the old couple strolling arm-in-arm.

Then there is the young mom pushing a stroller and dragging a reluctant toddler along in her attempt to get here she needs to be. Again a very different message from the young guy wearing head phones stomping down the street.

We all exhibit different styles of walking depending on our mood and intention without really giving it much thought. But as you can see there is much more to this than putting one foot in front of the other.

So how can we use this knowledge to silently communicate to our horse our mood and intention? Simple - all you have to do is be aware of your intention and your feet will automatically follow.

Let's imagine you are going out into the field to catch your horse.

Scenario #1 You were stuck in traffic and arrived at the barn half an hour later than you wanted to, which means you will have to rush through your chores in able to have enough time to ride your horse.

You are hot, frustrated and in a rush, so with a halter over your arm you stride out into the field to catch your horse - you have no time to dilly dally - you are on a mission.

Scenario #2 You have finished your chores, sat for half an hour on a straw bail in the sunshine enjoying a picnic lunch with your friend. You then decide to go for a trail ride so armed with a halter you head out to the field to catch your horse. As you wander down the field towards him you take time to stroke another horse who ambles up to you for some attention.

Scenario #3 You are in a really good mood - you have two weeks off work and are excited to start your holiday with a trail ride. You are in holiday mode, you have your favorite tunes playing on your phone, the sun is shining, life couldn't be better. So armed with a halter you walk briskly towards your horse, swaying slightly to the beat of your music.

All of these scenarios are totally believable and now you are starting to become aware of how our "walk" actually "talks". As you can imagine each different style of walking will send a very different message to the horse. Horses are excellent at reading our body language, especially our feet as they are the most honest part of our body. This is because we put the least amount of conscious thought into what our feet are doing.

The key to using your "walk to talk" is to become more and more aware of your own emotions and intentions and how that is manifested through your body language and silent communication. As you change your mood and your intention your style of walking will change automatically.

How to tell if your Horse is Stressed Out

Stress effects all of us and is caused by a myriad of events and circumstances.
Any of the following can cause you to feel stressed:
over work – busy schedule – peer pressure – rocky relationships – worrying about the worst-case-scenario and the list goes on.
People cope with stress in a variety of ways, one of which is:
Repetitive Behavior
Repetitive behaviors can help to soothe, pacify and reduce that anxious feeling of dread caused by stress. Some are innocent with no obvious side effects like foot bouncing, strumming your fingers, twirling a strand of hair or even nail biting. Other coping strategies can be self destructive like smoking, heavy drinking or binge eating.
Humans have developed a wide variety of diversions and hobbies that we find relaxing because of their repetitive nature. Activities like knitting, yoga, golf (sorry to all the golfers out there) to watching the same TV show over and over. The repetition provides a predictability which helps us feel in control.

Horses can also suffer from the overwhelming feeling of stress. It is sad to say most of their stressful situations are caused by humans. Just like humans, they too, have devised a variety of ways to cope with their misery. The name given to these behaviors is "Stable Vices".
It is a terrible description, and grossly inaccurate. The horse is not being deliberately naughty or wicked. He is simply using whatever means he has at his disposal to communicate to his owner that there is a serious problem. His body language is very often loud and extreme in his attempt to communicate to his owner that he is feeling stressed and extremely upset with his situation. In this chapter we are going to look at how these behaviors manifest.

Most equine stress relievers are self-destructive so we will also look at how they damage the horse's health. In conclusion we will talk about how to remove the cause and hopefully allow the horses to live a more relaxed and happy life.

We will start with the most common scenario and progress to the most self-destructive but also least common stable vice. When horses are confined either in a stable or paddock with insufficient companionship, exercise or very little grazing time they become stressed. Some people like to put a gentler name to it and call it boredom. It is a common misconception among horse owners, who feel that if the horse is well fed, has access to clean drinking water and housed in a nice cozy stable all his needs are met.
But think about it. The solitary horse has no way of knowing if he is ever going to see another member of his own species and as days turn into weeks - months and perhaps years he becomes more and more worried. Remember horses are very social herd animals so to be confined and alone is the worst kind of punishment.
So boredom doesn't really describe what he must be feeling.
To help soothe his nerves and make himself feel better the horse may develop compulsive repetitive actions. Over time these actions can become so obsessive they establish biochemical pathways similar to drug addiction.

Cribbing and Wind sucking
The early signs to both cribbing and wind sucking are when you notice chewed edges to the stable door and surrounding area or chewed fences if the horse is kept in a small bare paddock or pen. This "vice" (and I really don't like that word), starts out as wood-chewing. Perhaps the horse is driven by an internal need other than just boredom, hunger or desire to escape his confines and find other horses.
By nature, horses are grazing animals, spending much of their day feeding. The grazing horse has a continual flow of grass mixed with saliva going into the digestive tract. This mash of chewed grass buffers stomach acid and enzymes, which are also continually being produced in the gut to break down the fibrous material.
If the horse is deprived of food for long periods of time (which could be only half a day) imagine the build up of bile over even an afternoon, with no food to soak it up. This build up of bile and stomach acids can cause internal problems. The most common problem is colic and ulcers.

But lets go back to the wood chewing

We know for a fact that wild horses self-medicate when they are not feeling good. Throughout my extensive field research I have not found many differences in the natural way of being between wild and domestic horses, other than humans have restricted and fashioned their behavior to suit their needs.

What if! When a horse has an empty stomach and those acids build up causing him discomfort, he is driven to eat something, "anything", to soak them up. If the only semi-edible material available is the wooden stable door well guess what! He is going to chew it.

You see why I am reluctant to call this a "vice".

But this is just the beginning as wood chewing can quickly turn into crib-biting and wind-sucking. Although they have separate names one is just the more advanced and extreme version of the other.

Crib-biting involves the horse grasping onto a fixed object, usually his stable door or a fence, with his front teeth and he sucks in large gulps of air producing a characteristic grunt.

Many years ago during my British Horse Society training, I was taught that these large gulps of air can bloat the horses stomach and cause colic, plus it probably doesn't do their teeth any good either.

One of my personal mandates when researching for this book was to "challenge" all the things I thought I knew, not necessarily throw them out the door but to ask different questions and look at them in a different light.

We know horses are smart and they have the ability to self-medicate which makes you ask, "why would they do something to make themselves feel worse?" There must be some pay-off to this obsessive behavior. More recent scientific research suggests that cribbing increases salivation and may reduce stomach discomfort.
However cribbing and wind sucking must give the horse some kind of comfort either emotionally or physically and it can easily become a habitual soother.
If we were to compare it to a human "stress soother" we could liken it to thumb-sucking. Most children who use thumb-sucking as a pacifier grow out of this habit. Unfortunately this is not the case with horses. Once the habit is established it seems almost addictive and the horse will grasp onto anything in order to gulp air even if they are turned out in pasture and allowed to graze 24/7.

Another habitual behavior horses develop when they spend long periods of time confined to a stable away from other horses and feeling "stressed" is weaving.

Weaving: When the horse rhythmically sways while standing still; it usually involves a swinging of the head and neck and a concurrent shifting of weight between the front feet. Typically the horse does this standing right at the stall door.
Note: Horses are predisposed to walk, graze and form social relationships. When these basic needs are denied they experience stress.
Weaving is thought to be associated with the urge to walk but develops into a habit that is internally rewarding to the horse.
I like to try and draw on human similarities, but I don't want you to think of this in the same light as someone with a mental illness that sits in a chair habitually rocking back and forth. The horse is not mentally ill.
I would describe weaving more as an alternative to dealing with the real problem. Similar to keeping busy doing something else, perhaps obsessive cleaning or procrastinating and doing lots of little jobs rather than tackling the job that is on your mind and really needs attention. The pay-off is you did achieve something and you were legitimately busy just like the weaving horse. However the real problem remains and is still on very much on your mind.

A variation to weaving is
Stall Walking or Box Walking.
Just like the name suggests the horse circles compulsively in its stall. He may move fast or he may simply walk normally. He may also walk back and forth from wall to wall. He is obviously distressed and it is equally distressing to watch. This behavior burns calories and can lead to weight loss.
You may have seen similar behavior with caged animals at the zoo.
Please don't take this the wrong way. I am not suggesting keeping horses in stables is cruel or will result in them feeling so stressed they develop compulsive behaviors or what the scientists like to call "Stereotypies".
But we do need to ensure horses get enough "down time" when they are allowed to simply graze and socialize with other horses in a pasture.

Another stable vice that can have painful outcomes, more for the handler than the horse is
Biting.
The stressed and anxious horse may reach out of his stable, ears pinned and teeth bared to bite at passers-by, human or other equines. The aggressive behavior may be more exaggerated at feed time or perhaps when he is being groomed. This extreme behavior is his only way to get your attention and tell you something is really worrying him. The horse may be claustrophobic or feel very uncomfortable with the horse he is next to, or if the behavior is worse when you are grooming him, it may be he is in pain and requires more gentle grooming. Horses usually have a reason for acting the way they do. One thing is for sure - it's not about you. Biting is one of the most dangerous (for their handlers) habits that a horse can develop. Once a horse has made the decision to bite you, there is no escape. If he wants to bite you, consider yourself bitten. So be aware of the signs and make sure you keep a safe distance.

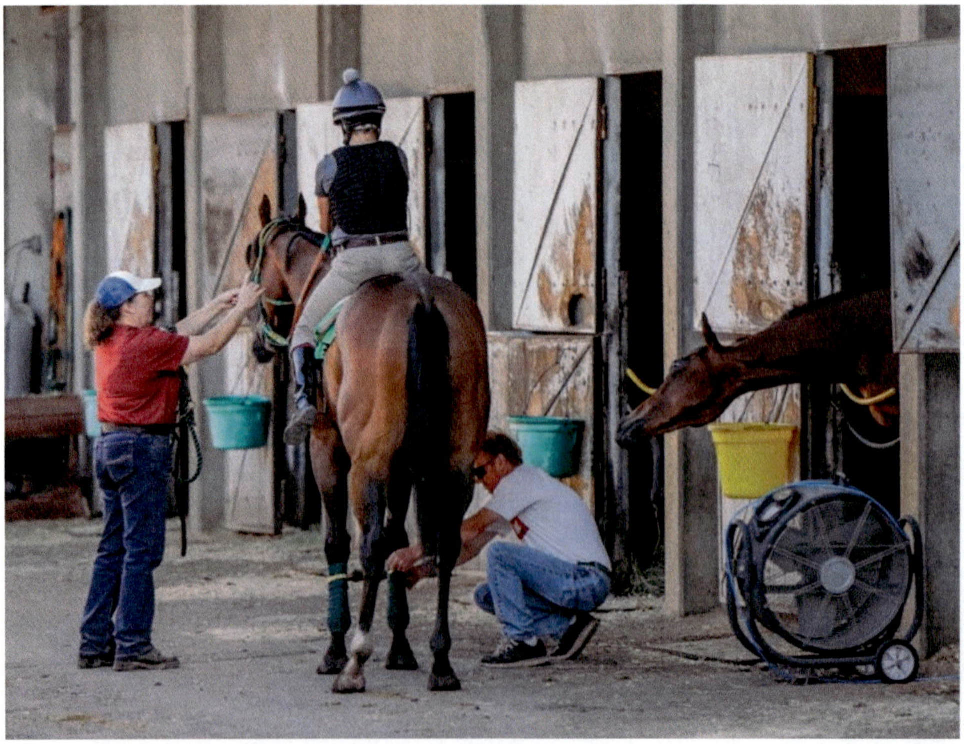

This form of body language is very common in the row stabling at the racetrack. The highly-strung, young (2 year old) Thoroughbreds, were well fed and cared for. However during the racing season, they are stabled for 5 - 6 months with no down time in the pasture to graze and socialize with other horses.

Banging the Stable Door and Kicking the Stall Walls.
This body language or habit is really hard on your nerves and your facility.
The horse will continually paw and bang on the stable door with his front leg and hoof to get attention. It usually works; sometimes he attracts negative attention and someone will yell at him. Other times his irritating behavior pays off and he will get an extra slice of hay to shut him up. But even when he doesn't get human attention the rhythmic banging can give him some kind of soothing effect. His basic needs are food, companionship and exercise. When stabled he has no way of knowing when any of these needs will be met and feels helpless. Pawing and banging may help him relieve that stress.
Similar to drumming your fingers on a table when you are waiting for a desired outcome, the rhythmic sound seems to help.

Some horses are **Stall Kickers**. They stand with their hindquarters near a wall and rhythmically thump the wall with one hind foot while the head bobs in a reciprocating motion. Again we can only assume that this action brings some kind of stress relief.
Others will kick at the wall with both barrels. They may give 2 or 3 hard kicks in a row. Because of the energy and balance required, this double-barreled kicker doesn't keep going like the one-legged kicker. The explosive kicker can wipe out a stall wall in a single kicking bout to say nothing of the damage that can be done to his hind legs.
Kicking is part of the socially acceptable play among horses, so stall kicking might begin as a natural behavior, but if the horse finds some kind of reward or pay-off it quickly becomes an obsessive habit.

Here is the really bad news: many stable vices, like door banging and stall kicking can be socially contagious. Neighboring horses might merely mimic the behavior of the kicker or might get involved in kicking as an interactive game. Nervous or fearful horses may interpret the sound of kicking of a nearby horse as a threat and will kick in defense. However, stall kicking is not usually an act of aggression toward another horse, but you can't rule out personality disputes. Kicking does occur between neighbors who don't get along.

So what can be done?
Interpreting the body language is the key to understanding why horses do the things they do (there is usually a reason for their actions). Once you recognize the action for what it really, is "a plea for help", you can simply remove the cause.
You can provide more pasture time, exercise or perhaps a new neighbor.
Some owners provide the horse with a distraction like a stable toy.

The most disturbing manifestation of stress is fortunately also the least common.

Self-Mutilation

This kind of destructive painful behavior is not confined to humans.
It's commonly is called flank biting or flank sucking.
The biting is one aspect of a cluster of behaviors classified as self-mutilation and the horse likely will incur serious self-injury during these explosive episodes. It is extremely violent and scary to watch. The horse will bite at his flanks in a seemingly uncontrollable violent outburst.
Typically spinning in circles, bucking, and kicking out with one or both back legs while nipping at the flank, shoulders, or chest. In extreme cases, the horse can violently lunge its body or head into a wall or other solid object. More rarely, a horse might "throw itself" to the ground. A single episode can last from a few seconds to several minutes of uninterrupted self- abuse. The horse will be so wound-up and excited he will work up a lathery sweat.
The total daily time spent self-mutilating can vary from a few seconds to an hour or more. In addition to bite wounds, the horse can injure his legs and feet from the spinning and kicking.

Self-mutilation can be seen in many different species, including humans.
Humans have devised all kinds of ways to harm themselves as a diversion from addressing the real problem, from pulling out their hair, biting their fingernails or lips, frantically scratching themselves, or deliberately inflicting burns, cuts, or other wounds.

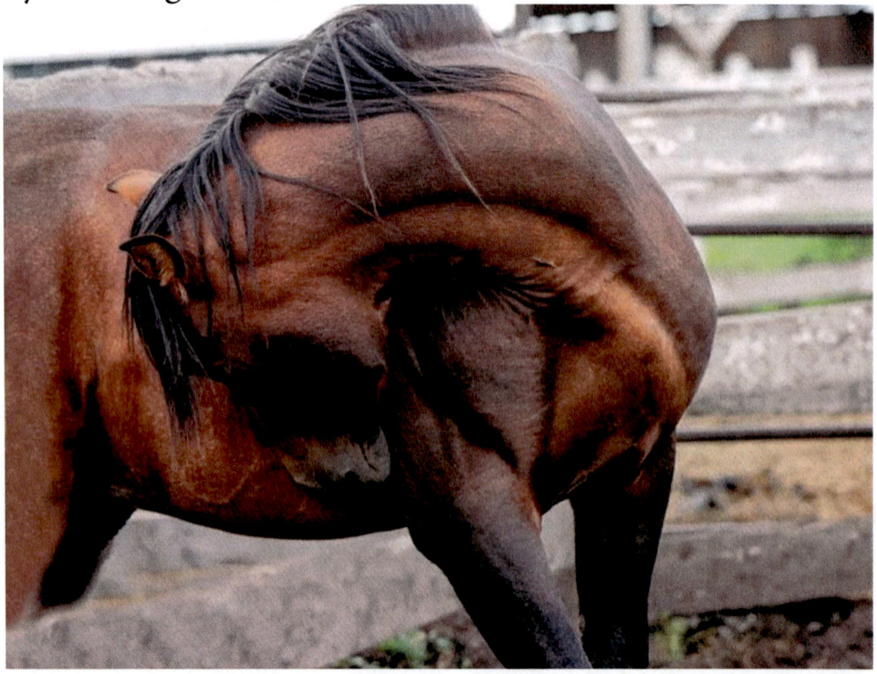

Bonus Chapter

Herbs that Heal Horses

While researching for this book I spent several weeks in the Rocky Mountain foothills studying the wild horses who inhabit that area. There are many theories as to where these horses came from. Some feel they are descendants of the pit ponies who were released from the coal mines once production of coal ceased. Rumor has it that these ponies were released into the wild to fend for themselves. Others think they are of mustang descent and were brought to the area by native tribes hundreds of years ago. The government insists they are feral (unwanted horses) abandoned by owners who no longer want them. Whichever story you choose to believe, the truth is, there are bands of wild horses not only surviving but thriving in the Alberta foothills.

These horses can teach us a lot about how to live a healthy lifestyle.
The secret of their success is fresh air, a plant-based diet and lots of exercise.

OK, fresh air is a given, and they move from place to place in search of food and avoiding predators. This leaves the real clue to their success as a plant based diet — notice I didn't say grass. These horses roam free on hundreds of acres of rolling hills and bush, which means they have free access to a wide variety of plants, herbs and wild grasses. It is this variety that keeps them healthy. They seem to have an intrinsic knowledge of which plants act as medicine when they are not feeling well. They have the inherent ability and the means to use natural healing herbs and plants to self-medicate.
In this chapter we will look at some of those plants and their medicinal properties.

You may be asking….. "Well that is all very interesting but what has that got to do with my domestic horse?"

The Good News: All horses have this ability when given free access to the right plants and herbs. They will only seek out and eat the grasses and herbs that will heal them (unless they are deprived of adequate nutritious food). They do not over medicate or over eat them when it is not necessary so develop immunity to the healing properties.

The Bad News: Over the years land owners have desperately tried to eradicate any "weeds" from their manicured paddocks. They have fenced the fields and pastures with post and rail fencing rather than hedgerows and trees. Most domestic horses eat exactly what their owners give them which is usually a very limited diet of grass (if they are lucky) and processed foods. Many are kept in stalls or small turn-out pens for long periods of time and perhaps exercised for an hour a day. If it fits their owner's schedule, some horses only get ridden on the weekends.

Two of the vital ingredients to a healthy lifestyle have been eradicated - just leaving fresh air. No wonder horses' owners are spending so much money with the veterinarian, just to maintain their horses health.

The Good News: There are simple actions that you can do to provide your horse with exactly what he needs.

Take your horse on a little nature walk - look for hedgerows that have not been weeded, manicured or treated with pesticides and allow your horse to graze on a loose rein at will. Pay close attention to what he chooses to eat and what weeds or shrubs he avoids. Take notes - does he just eat the leaves or does he burrow into the soil to get at the roots?

There is a behavioral science study in the UK called zoopharmaconosy that conducts experiments with horses who are experiencing various ailments. They take them on hedgerow walks and observe which herbs and plants the horses choose to eat - then they analyze any change in their condition. This is just a brief description of the scientific work they do. If you would like to learn more about their research do a Google search for zoopharmacognosy.

We will take a look some of the most common "healing herbs". There, of course are many more that you can research independently. It is also important to note not all plants / trees and grasses are suitable for horses, in fact some are extremely toxic. As a responsible horse owner it is imperative that you know what these are and make sure they are not growing in or lining your horse pastures.

Mint

Mint has an antispasmodic effect on the digestive system making it a natural solution for horses with tummy ache. Mint soothes the digestive tract and helps to expel gas. Mint also has anti-inflammatory properties and helps to reduce joint pain and swelling. An added bonus is it gives them nice fresh minty breath.

Rosemary

This herb has anti-inflammatory properties which helps reduce swelling and relieve pain. Rosemary leaves have an internal vermifuge effect and will come through the pores of the skin to make the horse less appealing to external parasites. It has a pungent smell which may deter horses from eating it.

Parsley

This is natural diuretic (promotes production of urine). It also works on the adrenal glands, is beneficial for optic and brain nerves, as well as the whole sympathetic nervous system.
It can benefit horses with coughs and arthritis, too.

Chamomile
This is an anti-inflammatory and analgesic (pain-killer) which acts as a sedative and relaxant.

Thyme
Thyme is very beneficial for the lungs and is particularly useful for coughs. It is also helpful for digestive issues.

Wild Brambles
Horses will eat all parts of the plant, thistles and all. They particularly enjoy the berries which provide a good source of vitamin C.

Milk Thistles
Horses find the purple flowers of the milk thistle yummy and they will eat the whole plant thistles and all. Of course they do prefer the younger stems. The seeds of the milk thistle are good for the liver.

Wild Garlic
Wild garlic acts on bacteria, viruses and intestinal parasites and has been used medically for over 5000 years. It boosts immune function, aids digestion and respiratory systems.
Wild garlic helps to normalize blood pressure and reduces fever.

Rose hips
Rose hips are the berry-like fruits of the rose bush which are left behind after the bloom has died. They are typically red or orange.
Rose hips contain high quantities of vitamin C, iron, copper and biotin they are excellent blood cleansers. They are also good for the kidneys, liver, adrenal glands and stimulate the immune system.

Dandelions
Horses love dandelions which is good news as the leaves and roots are especially good for their liver and kidneys.
Dandelions can act as a mild laxative.

Common Stinging Nettle
Horses seem to know that nettles are rich in iron and vitamin C. They also aid the circulation and cleanse the blood making them especially beneficial for horses suffering from laminitis and arthritis. Nettles provide horses with a natural "spring tonic".

Clover
Clover provides a great source of energy, protein and fiber for horses. However new spring clover is very rich and may aggravate horses prone to laminitis and founder. The early growth stage in clover contains high amounts of soluble sugars. These sugars decrease as the plant matures later in the season.

Horse Watching 101

Horse watching is a fascinating hobby. The more you participate, the more you will come to recognize how horses communicate to each other and to us.

The secret key that unlocks this fascinating world of equine behavior is to quietly meet horses in their own space and share the moment. Allow yourself lots of time - this is something you can't rush. Push aside all the things you think you know, have an open mind and no hidden agenda. If all you learn from the experience is the ability to quiet your spirit and unplug from technology, it will have been time well spent.

Top Tips for "New Horse Watchers"

Always get permission from the horse owners to observe and photograph their horses.

Never just hop a fence into a field full of horses (they belong to someone and you will be trespassing on private property).

Wear sensible clothing, sturdy boots, long pants and hat. You will be out in the country so bug spray is a must.

Binoculars, camera and a note pad is all the equipment you will need. It is important to watch the horses from a distance because your presence will alter their behavior.

Schedule enough time – this is something you can't rush.

Take bottled water – stay hydrated, but be sure to take all your litter with you when you leave.

Join a group of other Horse Watchers. It's always fun to compare notes and point things out to each other.

Record your sightings and findings – keep a journal.

Share your insights and experience on The Horse Watcher Facebook page.

Equine Experiments

The research behind the discoveries

Most of the research for this book was "field research" which literally meant sitting hour after hour in pastures with horses observing, documenting and photographing their behavior and body language. "Horse Watching" is a pastime I would highly recommend. When you leave all your personal worries and baggage at the gate and are quietly present with the horses you will be amazed at how much you will see. This process provided me with amazing insights into their personal world, but every now and then questions would arise that their normal day-to-day activities just did not address.

Questions like:
Can horses count?
Do horses have a favorite color?
What aromas are horses either attracted to or repulsed by?
Can horses problem-solve?
Do horses sense our energy - and how could you prove it?

So with the help of some friends and their horses we conducted a variety of little experiments. We also staged some unusual situations to observe how the horses would react and interact in order to answer some of those puzzling questions.

Some of these "experiments" successfully answered my questions, some experiments uncovered even more questions and some were just a bust. Despite being a little unorthodox, they were lots of fun for everyone including the horses. This is why I have included details of how the experiments were set up and some of the results.

The experiments are fun for both humans and horses (many of them include food) and easy to re-create with your own horses. If you get different or unusual results I would love to hear from you. There is a Facebook group called "The Horse Watcher" where you can share your experiences and results.

PS. If you are a riding instructor that offers summer camps, many of these experiments would provide a fun activity plus provide an excellent introduction into equine body language and behavioral studies.

Can horses count?

The objective was to determine if horses could not only count how many pieces of carrot were dropped into a bucket but would choose the bucket containing most carrot treats.

We tested 60 horses with this experiment, mainly because it was fun and everyone at the barns we visited wanted to put their horse to the test to prove how smart they were.

What you need: Two identical buckets and a bag of carrots (cut into bite sized pieces).

The horse was positioned between the two buckets (a little way back) so he could see what I was doing.

First I gave the horse a piece of carrot so he knew what the treat was.

Then I dropped one carrot in the first bucket and dropped three carrots in the other bucket.
They were dropped separately so the horse could hear each carrot landing in the bucket.
The horse was allowed to walk forward and choose which bucket he wanted.
Would he connect the sound of three carrots falling into the bucket plop-plop-plop as more rewarding than just one plop of the first bucket?
Would the horse choose the bucket with more carrots in it?
Was he able to count?

We alternated which bucket had more carrots in it and each horse was given three tries.

Results:
Of the 60 horses we tested 50% totally got it. They watched attentively and without hesitation went to the bucket with three pieces of carrot every time.
35% of the horses went to the same bucket each time - I don't want to say these horses are not smart, in fact they are probably very easy to train. After all, by repeating the same action (same bucket) they got a reward, which is in fact, how we train horses.
15% of the horses thought the experiment was dumb and either randomly selected a bucket each time or chose not to participate at all. Some horses just turned around and walked off.

How do foals learn what is good to eat?

Think back to when you were little. Do you remember your mother telling you to eat your greens, they are good for you? Or when you reached for something off the floor your mom would stop you from putting it in your mouth, saying "Don't eat that - it's dirty."

Do mares do the same kind of thing with their foals? The objective was to see if the mare would move her foal away from the food that was not good to eat.
So I decided to put it to the test.

What you need:
Three buckets - one containing oats, one containing apples, the last one containing cat kibble.

We placed the mare and foal in a safe enclosed area, then set out the buckets of food.
Note: The bucket containing cat kibble was in the middle.

Results:
The mare first ate the oats, she passed by the bucket contains cat kibble and continued on to eat the bucket of apples. Her foal paid close attention to what momma was doing.

Knowing how inquisitive foals can be I half expected him to go over to sniff the cat kibble.
At that point I wondered if his mama would chase him away.

Although not the result I expected it did prove that mares teach their foals by example, do as I do, unlike the human method of do as I say.

Do bugs bug him?

This may or may not work as the bugs are fake - that's why it is called an experiment.

What you need:
Large plastic bugs or snakes (we bought ours at the dollar store).
The fake wasps were bright yellow, a color usually associated with danger.
A shallow feed bowl and some oats.

Large plastic wasps were arranged in a shallow feed bowl along with some oats.

After calling the horses over to the bowl — we stepped away to observe their reaction and body language.

We tested this experiment with several groups of horses. The results were the same every time. Although the horses approached with caution, they were not fooled by the fake bugs and most of the horses just ate around the bugs. One horse picked up the bugs and tossed them out of the bucket.

Toy horse experiment

What you need:
A large toy horse and a camera as you will want to take pictures of the first encounter so you can study the horses' micro-expressions later.

If you enjoy Horse Watching, you will really find this little experiment amusing.
Objective: To discover if horses recognize the toy as a horse, and if so, would they try to make friends with it or reject this strange new addition to their herd.
The first encounter may be confusing to the horse because even though the toy looks like a horse it does not act like a horse. Of course the toy is unable to respond to the horses' body language (nonverbal communication) be it friendly or aggressive.

There was no right or wrong result for this experiment. However it was a wonderful way to observe equine body language and facial expressions.

As you look through the sample photos from our experiment, try to spot tell-tale body language clues.

Notice the low (non-threatening) position of the horses' heads and necks, ears pricked forward and soft, kind expression in their eyes. Also notice how the horses are carrying their tails. All three horses are allowing their tails to hang soft and relaxed.

The brood mares were very interested in this “new addition” and sniffed it from head to tail.

They then put their heads together as if to discuss what to do with it.

The stallion was totally freaked out and would not go anywhere near the toy horse

We introduced the toy horse to a couple of llamas, but who knows what llamas think.

What fragrances are horses attracted to or repulsed by?
And why as horse owners should we care?

For centuries people have used essential oils and aromatherapy for general health and emotional well being. We are attracted to certain scents and when inhaled they trigger an emotional and physical response.

Horses have a keen sense of smell so by understanding which scents they are attracted to or repulsed by may help:
Soothe and calm a nervous horse.
Help you to bond with him (his favorite aroma might become your favorite perfume).
Put a horse at ease in a stressful situation like a new stable or trailer.
This is not an exact science and I am sure not all horses are attracted to the same smells, but I am sure it would be beneficial to know a little more about the preferences of your own horse.
What you need:
A variety of essential oils - clean cloths to apply the oil to.

We used Lavender, Peppermint, Eucalyptus, Ylang Ylang, Green Apple, Jasmine and Anise.

The experiment:
The experiment took place in an indoor arena - we created a large circle using upturned buckets. On top of each bucket we placed a cloth containing a few drops of essential oil.
The horse was led to each bucket on a loose rein and allowed to sniff the fragrance.

We paid close attention to his body language, paying particular attention to his ears, eyes, nose and lips. We also took into account whether he wanted to move closer to the bucket or away from the bucket.
We gave the horses' response to each fragrance a rating of 1 to 10.
1 - Strongly dislike (tried to get away as soon as possible.) 10 - loved it so much he was trying to eat the cloth.

We tested 35 horses (different breeds, sex and age).
The Results:
This experiment had a clear winner. The favorite fragrance with every single horse was Anise closely followed by Peppermint.
A really surprising result among the horses we tested was their strong dislike for the floral fragrances Jasmine and Lavender. They would take one whiff and turn their head and neck in the opposite direction, their body language speaking volumes.

Do horses have a favourite color?

Just to be clear this experiment has nothing to do with whether your horse prefers a blue saddle pad and a pink halter. The purpose is to discover if he is more sensitive to or worried about certain colors.

What you need:
Five buckets all the same size but all different colors. A big bag of carrots chopped into bite sized pieces.

experiment:
We placed all the buckets evenly spaced in a row. We brought the horse into the arena and stood him about 15 feet away from the buckets.
I walked up to the horse and gave him a piece of carrot so he knew what the treat was. Then (with him watching) dropped one piece of carrot into each bucket.

The horse was allowed to walk towards the bucket of his choice.
The buckets were then re-arranged (again with the horse looking on - we were not trying to trick him).
We repeated the experiment.
There were five different colors so we repeated the process with each horse five times.
This was a fun game for the horses as they got lots of treats.

The object of the experiment was to determine if the horse had a favourite color.
We tested 35 different horses but none of the horses seemed to care what color bucket they ate out of.
Most horses kept going back to the same spot, regardless of what color bucket was placed there. From the horses' point of view the same action got the same reward.

Twinkle Toes Experiment

Do horses read our body language?

It is said that the feet are the most honest part of the human body. Our toes always point in the direction we really want to go. They also point towards people you are attracted to and away from those you don't really like. This means, if you are unsure if someone finds you interesting or they are just listening to you to appear polite - glance down at their feet to discover the truth. Fascinating facts, but let's revert our attention back to the horses.
The purpose of this experiment is to either prove or disprove that horses read our body language. The goal is to interpret how they react to what our feet are doing and silently saying.

The experiment is done with a single horse (any age / sex or breed) in a riding arena or bare paddock, so the horse is not distracted by grass.

It should take around 10 - 15 minutes and flows almost like a dance.

Begin by standing in front of the horse and engage his attention. Once you have made a connection with your horse begin the little dance.

The steps

1. Adopt the Superman pose feet apart (a very strong dominant stance) hold for few seconds as the horse decides on his reaction to this.
2. Lower your arms to your sides and start walking slowly backwards - if the horse follows you speed up until you are almost running backwards (try not to fall over).
3. Stop.
4. Point one foot towards the gate (or out to one side).
5. Bring it back (feet together).
6. Tap your foot impatiently.
7. Turn around and walk slowly away from the horse - if he follows you start running (again try not to fall over).
8. Stop.
9. Turn around to face the horse - now do a little dance of indecision - pretend to step one way, then change your mind and switch direction. Maybe spin a little (as though you really can't make up your mind where to go next).
10. Walk up to the horse and give him a pat on the neck.

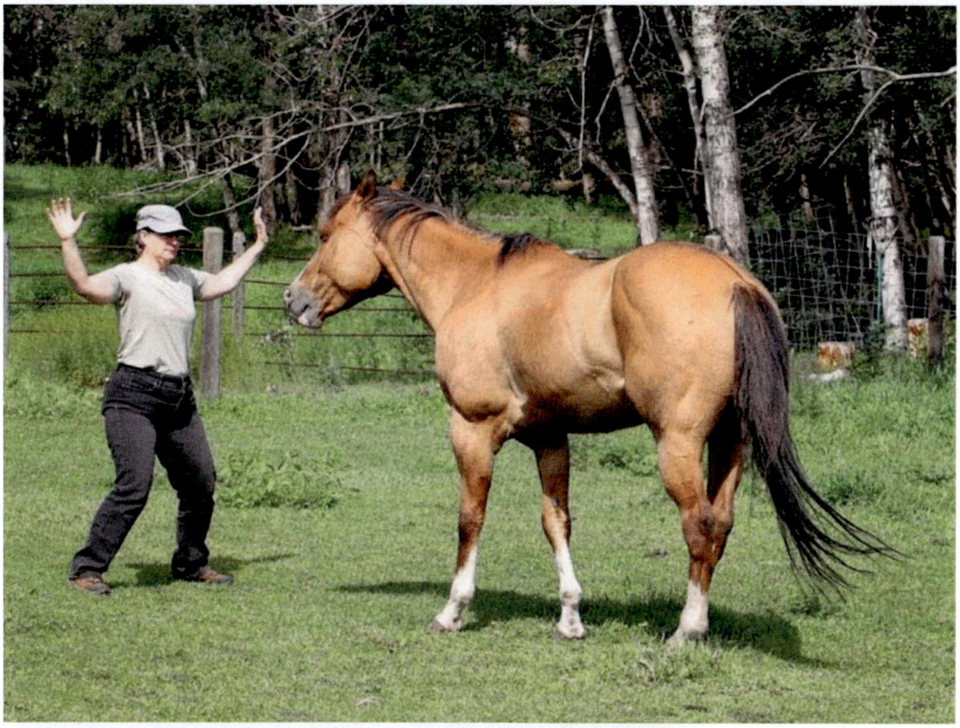

Catherine has adopted a very dominant pose - her head is held high her arms also held high with palms facing the horse. Her legs are apart taking up as much personal space as possible.

What can you determine from the horses nonverbal response to her stance?

The horse's ears are pinned, eyes closed, nose jutting forward and tail swishing. He does not seem happy.

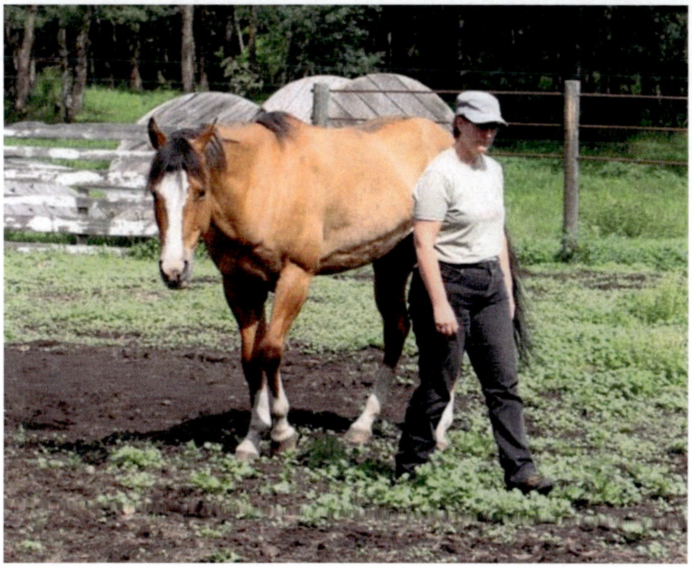

Now his owner has taken a softer stance the horse is happy to follow her.

As Catherine does her little dance of indecision notice how the horse has spread his front legs ready to go in whichever direction his owner finally decides to go.

The horse definitely read Catherine's intentions by paying attention to her feet.

These were all normal human foot movements. Now try mimicking equine foot movements and see how the horse reacts to or interprets those.

1 - Stamp your foot - try once then stamp it a few times in succession.

2 - Paw the ground (impatient).

3 - Jump on the spot (anxious).

4 - Rest one leg (relaxed).

5 - Cross one leg over the other (this shows you are not going anywhere).

Remain silent throughout the exercise. Do not give the horse any verbal cues.
There are no right or wrong outcomes, this is not a training exercise it is designed to quietly observe how in-tune the horse is to your body language.

Catherine stamped her foot, but turned her body away from the horse (her frustration was not aimed directly at him). The horse seems to be looking around to see what could possibly be upsetting his owner.

Catherine pawed the ground.
The horse seemed very interested in this and lowered his head and neck so he could get a better look at her feet, but he did not mirror her actions.

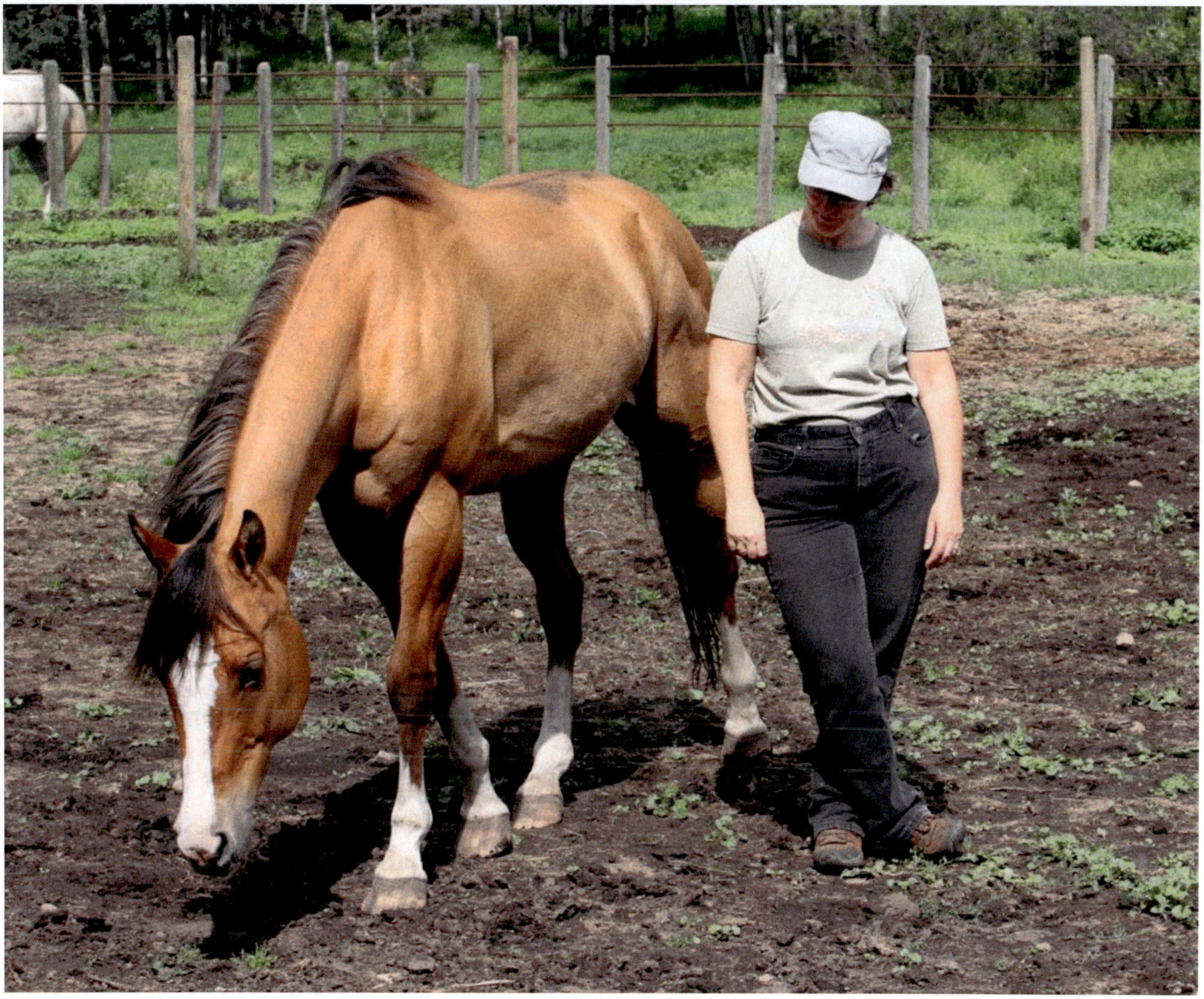

Finally Catherine crossed one leg over the other indicating she was not going anywhere. The horse visibly relaxes and lowers his head.

If you do this little experiment with your own horse, persuade a friend to take pictures as you go through all the different actions and steps so you an analyze the results later.

Are horses into to our "energy"?

There is a common myth bandied about that horses can sense our mood or energy, they somehow can tell if we are excited / happy / sad or scared.
I personally believe that it is our body language that gives away our emotion. Because it is a popular belief my challenge was how to prove this, and also to photograph it.

The objective: To observe and understand how horses react to both high energy (excitement) and low energy (calmness).

Equipment: Small table / playing cards / duct tape / blanket and a book.
Best performed in a small field with a mixed herd of mares and geldings.

Experiment:

1.One person will slowly walk into the field, place a blanket on the ground, sit down and read a book, emitting low energy and a feeling of calmness.

Observe how the horses react.

2. In the next part of the experiment two people will play a high energy game of cards -Snap. Two people carry a small card table into the field and place it near where the horses are grazing. Next duct tape is placed over their mouths so they can't accidentally yell "Snap."
They then proceed with their game (the players should totally ignore the horses, definitely no eye contact). The players are encouraged to be very animated and slap the cards down hard on the table.

Observe how the horses react to this high energy scenario.

Results: Inconclusive

Elixis sat quietly on a blanket reading a book - the horses were very curious and came over to check out what she was doing.

Within a few moments the horses lost interest and went about their business.

You could deduce that because the human was calm and relaxed, the horses too felt calm and relaxed, but it's impossible to know for sure.

Melissa and Elexis get ready to start the experiment, duct tape in place over their mouths and cards at the ready.

Even though they play a vigorous game of Snap - the horses seem disinterested in their activity.

There is little evidence that the horses "felt" the girls energy. They paid little or no attention to the humans.

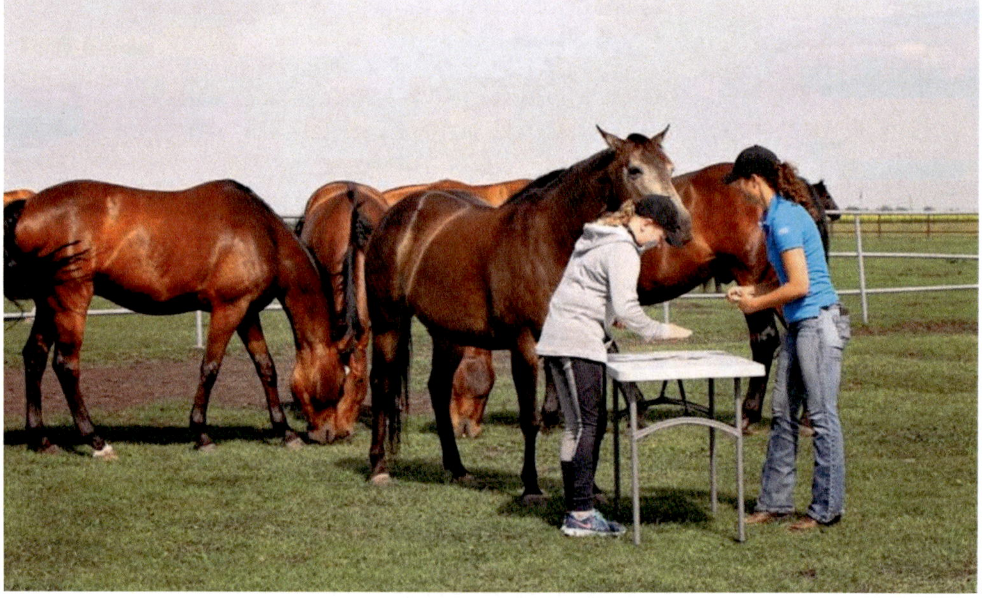

This experiment was carried out six times with different horses. Each time the results were the same.

Perhaps I need to develop a different experiment to answer the question:
Do horses feel our energy?

Problem Solving Experiments

These little "find the food" puzzles were originally designed to test the intelligence and trainability of dogs. They have been adapted to work with horses although the purpose is less to do with problem solving (we know horses are smart and trainable) and more as a fun way to study body language.

The tests can be done with a single horse or multiple horses at the same time.

As you observe do not concern yourself with whether or not the horses do what you expect them to do or how long it takes for them to solve the puzzle. Rather pay close attention to answering the following questions:

Q. Do the horses help each other solve the problem?
Q. If one horse solves the problem will he share the rewards with the other horses?
Q. Does one horse stand out as being smarter than the rest?
Q. Does one horse stay back and wait for someone else to solve the problem?
Q. How quickly do the horses solve the problem?
Q. Do they lose interest and wander off? Is that because they are not motivated by food (maybe their bellies are full) or do they just choose not to participate?
Q. Do the horses become frustrated with the challenge?
Q. Do the horses squabble and fight between themselves?
Q. Does the common problem help to bond the horses in a common quest?

Magic Buckets
You will need 3 identical light weight buckets and several pieces of carrot.
First give the horse a piece of carrot so he knows what he will be looking for - allow him to see you place a few more pieces of carrot under a bucket.
Step back and observe.

If the horse knocks over the bucket and finds the treat - repeat. This time add another bucket.

Not surprisingly, all the horses knocked over the bucket to reveal the treats very quickly. What was surprising is, when we tested multiple horses at the same time they quietly cooperated with each other, amicably sharing the treats and the game.

Behind the Scenes

Set up a solid barrier - give the horse a piece of carrot so he knows what the treat is. Stand in front of the barrier and drop a piece of carrot on the floor (make sure the horse is watching you.) Whilst he is eating go behind the barrier - attract the horse's attention. When he is looking at you show him the carrot and drop it on the floor behind the barrier - then step away and observe.

Will the horse go around the barrier to retrieve the carrot?

This test was almost too easy; it was hard to prevent the horses from following us as we went behind the barrier. The horses knew we had more treats so they were eager to follow us anywhere.

Food Under Towel

A small washcloth is placed over a treat (a quarter of an apple is best because the horse can smell and feel it). First give the horse a piece of apple so he knows what the treat is. Allow the horse to observe you placing the apple under the washcloth then step back. He will be able to smell the apple but he not see it which means he will have to switch sensory modes from visual to olfactory, making this game a little more challenging.

Keep a careful eye on what the horse is doing in case the horse tries to eat the whole setup, washcloth and all.

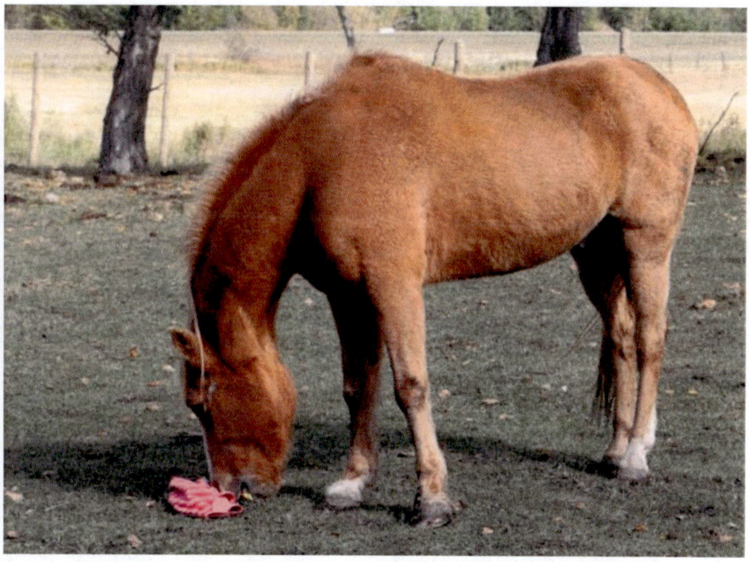

Remember the purpose of these little experiments is to observe the body language as the horses solve the puzzle.

Top Shelf Problem

Give the horse a piece of carrot then place more carrots on a high place - so the horse has to stand on something to get at the treat. Be sure the platform he is standing on will bear his weight.

Under Barrier

Make a ledge that can be used to hide a treat by placing a light piece of plywood (about 3-4' by 1.5-2') on a couple of tires (or other sturdy support about 8 inches high). Allow the horse to see you place half an apple under the ledge close to the edge, then let it sniff the treat. Will the horse use his nose or his hoof to flip the barrier or just give up?

Apple Bobbing (tree test)
You will need a whole apple with a hole bored through the middle and long piece of string. Tie the apple from a tree so the horse has to stretch to get it.
Lead the horse to the tree and show him the apple, then step back and observe.

Apple Bobbing (water bucket test)
A variation on this test would be to half fill a bucket with water and place the whole apple in the water. The easy solution to this problem would be to knock over the bucket, so the water and apple spill out making it easy for the horse to retrieve his treat, but will the horse figure that out?

The results:
This was by far the most challenging problem we gave the horses to solve. Single horses quickly became frustrated and gave up. However when two horses were allowed to work on the problem, they seemed to work together by trapping the apple between them. One horse could get a firm bite on the apple and retrieve the treat from the tree. When the apple fell to the ground both horses shared the fruits of their labor.

We did not restrict our research to just horses. Donkeys and mules were also invited to join in with the fun and games, after all they are equines too.

Are Horses Self-Aware?

Before we look at how to answer that question, consider for a moment the same question. Are you self-aware?
Regarding humans, self-awareness is having a clear perception of your personality, including strengths, weaknesses, thoughts, beliefs, motivation, and emotions. You know what you look like, although that may be a distorted view. You would be able to pick yourself out of a line-up.
You know what sex, height and body shape you are (again that may be a distorted view as self-image is very different from self-awareness).

You probably answered YES to that question. Well done, for the most part humans are extremely self-aware. Some take it to a whole new level and become self-obsessed but that is a discussion for another time.
But the question remains are horses self-aware and would they recognize themselves in a mirror?

Confession:
This burning question and the groundbreaking discoveries the experiment produced were part of research I conducted for an earlier book **The Horse Watcher** *Answers to questions you never knew you had.* The images recorded in that experiment provide amazing examples of equine body language. Although we will be searching for the same answer to the question our focus will be on how the horse non-verbally communicates his answer.

The experiment:
To ensure we had continuity between test subjects we only used stallions, between the ages of 3 and 12 yrs. who were typically housed in their own pen - away from other horses.

The objective was to determine if the horse recognized himself in the mirror (proving he had a good sense of what he looked like).
Bear in mind he has probably never seen his reflection before. Horses cannot see their reflection in water due to the positioning of their eyes and a small "blind spot" right in front of their face.
Although they may never have been exposed to a mirror before think how quickly small children figure out how mirrors work, and how fascinated with their own reflection they become. We know horses are extremely smart, so let's see if he can figure it out.

Step 1 A large Plexiglas mirror (unbreakable) was placed in his pen (the wrong way around) we encouraged the horse to go over and check out this new object in his pen.

The horse was curious about this new addition to his living space, he sniffed at it, but quickly lost interest.

Step 2 When the horse wandered away, the mirror was turned around and the horse was encouraged to go back and take a second look. This is when the fun and games really began.

This Andalusian stallion was extremely expressive making it easy to read his body language.

His initial response at the horse in the mirror was obvious surprise.

Surprise quickly turned into anger and aggression. The stallion used his front feet to pummel the strange horse in the mirror. He was annoyed that this horse had invaded his space. This behavior (pounding the mirror) went on for about 10 minutes - but of course his actions and aggression were mirrored back to him which meant neither horse was winning this battle.

So he lowered his head and neck in a submissive gesture and slowly backed away from the mirror.

He didn't got very far before deciding to spin around and have another go at chasing out the intruder (the horse in the mirror).

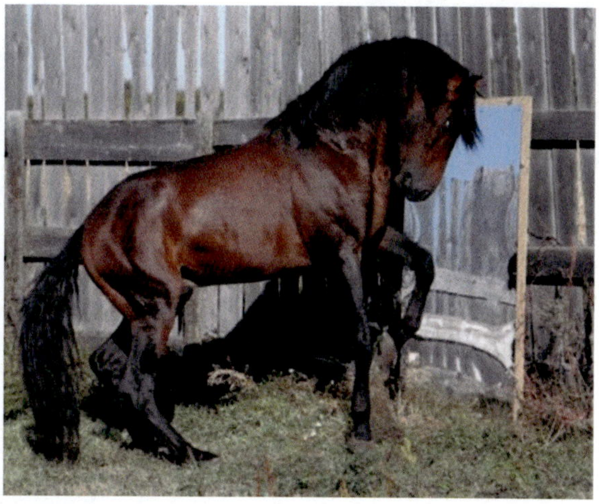

The horse was allowed to interact with his reflection and engage in his obvious inner turmoil. Although it was fascinating to watch we had to put an end to this behavior before he hurt himself. At that point the horse was chased away from the mirror and it was turned around again to the non-reflective side. The horse was eager to return to the mirror (or in his mind the intruder) only to discover the mystery horse was no longer there.

He looked over the fence.
He neighed and called out.
He looked behind the mirror.
He ran up and down the fence line frantically searching for the other horse.

It was evident that our stallion did not recognize himself in the mirror.

It was only fair to offer our test subject a reward for his explosive demonstration. The horse was taken into a stable and given a nice lunch of oats and carrots.

In the original experiment there was a second part to be conducted if any of the stallion owners thought their horse may have recognized himself in the mirror. After the horses had calmed down and had lunch the intention was to paint a large blaze on the face of the horse in either white or black finger paint. The purpose was to change his appearance. We would then repeat the experiment to see if his reaction was different now that his appearance had changed.

We had first conducted this experiment with 12 different stallions and each horse reacted in exactly the same way. That meant we never actually had to go to the next step.
I was personally a little relieved, because I never told the stallion owners prior to my visit that I intended to paint the faces of their prize stallions.
That part of the experiment was to be a surprise.

Conclusion:
Although the experiment did not either prove or disprove that horses have a level of self-awareness, it did prove that the 12 stallions we tested did not recognize themselves in a mirror.

Warning: If you try this experiment with your own horses be sure to use a Plexiglas mirror NOT a real mirror. Also keep a safe distance from your horse as his actions may be explosive and unpredictable. This is a whole new experience for him; he has never seen himself before.

More Books By *Linda Finstad*

The Horse Watcher

This beautifully illustrated book answers many of the questions you never knew you had, concerning Equine Behavior and Communication. "A must read for all horse lovers." (*On The Pony Club suggested reading list)*

How to Put Butts in Seats or Saddles

A practical "Hands On" guide for Coaches and Equine Clinicians who offer workshops or lesson programs. Never again worry about cancelling your clinic because no one registered.

Don't Shoot the Horses

Horses are capable of many facial expressions that range from the good, the bad to the downright scary. Linda has broken the unwritten rule among equine photographers of not only capturing those moments but publishing them in her book.
Warning - *when you open these pages you will be leaving your comfort zone.*

Available from: **www.TheHorseWatcher.com**

Who is Linda Finstad?

Linda Finstad is a Teacher, B.H.S.A.I., Entrepreneur, Professional Photographer, Author, lover of all things horse and devoted follower of Jesus Christ.

Equine Behavior Investigator

This may sound like a made up title.

However it perfectly describes what Linda has spent a lifetime doing in her quest to unravel the mysteries of equine behavior. Linda has spent countless hours observing, documenting and photographing both wild and domestic horses, to learn the answers to questions that have haunted her dreams. Questions such as why do horses seek a relationship with humans? We are a different species which adds nothing to their existence, yet they seek to please us. How do horses communicate empathy and understanding and how can we decode their silent conversations?

Linda doesn't give herself false airs by calling herself an expert or even a specialist. As she discovers the answer to one question another one seems to take its place. She humbly acknowledges that her research and investigations are on-going; there is always something new to learn and discover in the quest to truly understanding how horses communicate.

If you have the opportunity to watch a quadrille or musical ride you will realize that a sense of team work and cooperation is still evident in our domestic horses. In the case of the RCMP, sixteen horses perform elaborate maneuvers in extremely close contact. Each horse's success is very much dependant on the horse both in front and behind him.

Typically riding instructors teach their students to keep a "safe" distance from the horse in front and work independently, which is of course sound advice.

This is a quality we tend to overlook.

Human Equivalent

Although we do not think of ourselves as "herd" members we create lots of situations where our success very much depends on the cooperation and predictability of our team mates. A few that come to mind are: Synchronised swimming - Football - Mountain climbing.

INDEX

Made in United States
Orlando, FL
04 November 2026

90200815R00124